T0133971

Practical Guide to
Inspection, Testing and Certification
of Electrical Installations

Covers all your testing and inspection needs to help you pass your exams on City & Guilds 2391 and EAL 600/4338/6 and 600/4340/4 and Part P courses.

- Entirely up to date with the 18th Edition IET Wiring Regulations
- Step-by-step descriptions and photographs of the tests show exactly how to carry them out
- Completion of inspection and test certification and periodic reporting
- Fault finding techniques
- Testing 3 phase and single phase motors
- Supporting video footage of the tests contained in this book is available on the companion website

This book covers everything you need to learn about inspection and testing, with clear reference to the latest updates to the legal requirements and wiring regulations. It answers all of your questions on the basics of inspection and testing, using clear and easy to remember language, along with sample questions and scenarios as they will be encountered in the exams. Christopher Kitcher tells you what tests are needed and describes them in a step-by-step manner with the help of colour photographs and the accompanying website.

All of the theory required for passing the inspecting and testing element of all electrical installation qualifications along with the AM2, City & Guilds 2391 certificate and the EAL 600/4338/6 and 600/4340/4 qualifications is contained within this easy-to-follow guide – along with some top tips to help you pass the exam itself.

With a strong focus on the practical element of inspection and testing for NVQs or apprenticeships, this is also an ideal reference tool for experienced electricians and those working in allied industries on domestic and industrial installations.

www.routledge.com/cw/kitcher provides a large bank of helpful video demonstrations, multiple choice questions to test your learning, and further supporting materials.

Christopher Kitcher has worked in the electrical industry for over fifty years. He currently works as a freelance electrical installation lecturer and also as an NICEIC inspector for the Microgeneration Certification Scheme (MCS). Since retiring from the college environment he has been involved in bespoke training around the country for various organisations, and maintains his electrical skills by periodically working on site.

Practical Guide to

Inspection, Testing and Certification of Electrical Installations

Fifth Edition

Christopher Kitcher

Routledge
Taylor & Francis Group

LONDON AND NEW YORK

Fifth edition published 2019
by Routledge
2 Park Square, Milton Park, Abingdon, Oxon, OX14 4RN

and by Routledge
52 Vanderbilt Avenue, New York, NY 10017

Routledge is an imprint of the Taylor & Francis Group, an informa business

First edition published by Newnes 2007
Fourth edition published by Routledge 2016

British Library Cataloguing-in-Publication Data
A catalogue record for this book is available from the British Library

Library of Congress Cataloging-in-Publication Data
Names: Kitcher, Chris, author.
Title: Practical guide to inspection, testing and certification of electrical installations / Chris Kitcher.
Description: 5th edition. | Boca Raton : a Routledge title, part of the Taylor & Francis imprint, a member of the Taylor & Francis Group, the academic division of T&F Informa, plc, 2019. | Includes bibliographical references and index. |
Identifiers: LCCN 2018039247 (print) | LCCN 2018039496 (ebook) | ISBN 9780429869242 (ePub) | ISBN 9780429869259 (Adobe PDF) | ISBN 9780429869235 (Mobipocket) | ISBN 9781138613324 (pbk) | ISBN 9781138613331 (hbk) | ISBN 9780429462290 (ebk)
Subjects: LCSH: Buildings--Electric equipment--Inspection--Handbooks, manuals, etc.
Classification: LCC TK4001 (ebook) | LCC TK4001 .K48 2019 (print) | DDC 621.319/240288--dc23
LC record vailable at https://lccn.loc.gov/2018039247

ISBN: 978-1-138-61333-1 (hbk)
ISBN: 978-1-138-61332-4 (pbk)
ISBN: 978-0-429-46229-0 (ebk)

Typeset in 10/13pt Helvetica Neue by
Servis Filmsetting Ltd, Stockport, Cheshire

Visit the companion website: www.routledge.com/cw/kitcher

I would like to dedicate this book to my wife and my grandchildren for leaving me in peace when asked.

Contents

List of figures and tables

Figures

Tables

Acknowledgements

Central Sussex College, in particular Dean Wynter for allowing me unlimited use of the workshop facilities whilst taking photographs and filming.

Simon Wood of Megger UK for providing equipment and information whenever asked.

Dave Chewter for always being ready and happy to help with anything.

Phoebe, my granddaughter, for help with paper work.

Introduction

Inspecting and testing of electrical installations

We all use electricity every day and most of us just take it for granted that it is safe to use. Of course, for the majority of time it is. This is not usually down to luck, although when I think about some of the installations which I have seen over the years, I am well aware that on some occasions luck must have been around in abundance.

Over the years the way we deal with electrical installations has changed dramatically; this is of course down to education and experience. Apart from the use of modern materials and methods of installation we also have improved legislation in place which should ensure that all installations are inspected regularly.

When I first started full-time work back in the early 1960s there were massive house building projects being carried out all over the country, but testing and certification of new installations was virtually unheard of. When we completed a new domestic installation, the supply authority were really only interested in getting a signature from the person who was going to be expected to pay the electricity bill each quarter.

We used to do an insulation resistance test on the meter tails and the person who installed the meter usually did the same before connection, but that was all. The insulation resistance tester was not anywhere near as sophisticated as a modern one; we used to have to wind the handle of the instrument as it was a mini generator (Figure 1.1).

I remember clearly that if for some reason we had a fault due to a nail being driven through a cable, or some other fault which resulted in a bad reading, we would just remove the fuse wire from the rewirable fuses, or disconnect the neutral of the circuit concerned before the person arrived to install the meter. That way we could be sure that the installation would be connected and that we would have an electrical supply. It is usually easier to trace a fault if the system is live, particularly in the winter, as it is much easier to find a fault in a warm house with light than a cold house in the dark.

As far as earth fault loop impedance was concerned, the only time we measured that was when a survey was being carried out, and again the instrument was entirely different to the equipment used today (Figure 1.2).

All new houses had a copper or iron water main, as did most old ones. As you can imagine, the surface area of the metal from the water mains in contact with the soil was huge. This resulted in

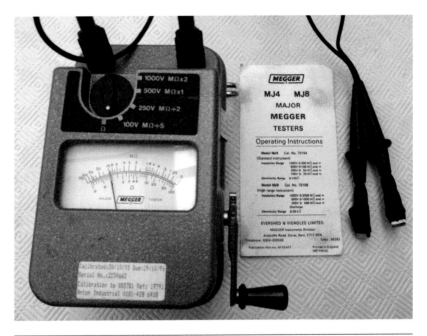

Figure 1.1 Wind up insulation resistance tester

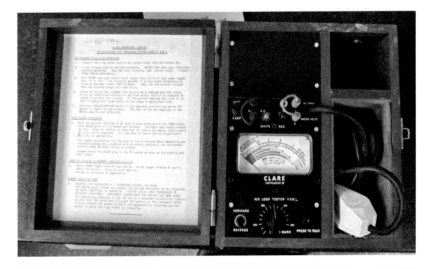

Figure 1.2 Earth fault loop impedance tester

very low earth fault loop resistance readings. This is because the resistance of soil is usually very low as there is such a lot of it.

As the years have passed more and more electrical equipment is being installed into buildings; it is also becoming more and more sophisticated of course. Health and safety, along with insurance, has also had a hand in making it important that in the event of a fault somebody can be held responsible. Usually this will be the person signing the document to say that the installation is compliant with the current edition of the wiring regulations (BS 7671). For this reason it is very important that we take the installation of electrical wiring, along with inspection, testing and certification, very seriously.

It is important that we not only know how to install all of our new fixed wiring correctly, but that we know how to verify and document it as well. Not only that! We should also be able to inspect an existing installation and, with the help of some testing where required, we should be able to verify that it is fit for continued safe use. Where damage or non-conformities are found, we must be able to identify them and make sound, professional recommendations about the installation.

We must also be able to relay this information to our clients in a professional, non-technical manner. Many of us will remember how difficult it was to understand the terms used in the electrical industry when we first started out.

Your client will need you to identify the technical detail, record it and then relay it to them in words which they can understand; of course before we can do that we need to understand it ourselves. Hopefully what follows in this book will be of help.

Video footage is also available on the companion website for this book.

CHAPTER 1

The legal requirements

Apart from the obvious safety reasons, we also have to concern ourselves with the legal requirements for electrical installations. The main statutory documents which we need to comply with are:

* The Health and Safety at Work Act 1974 (HASWA)
* The Electricity at Work Regulations 1989 (EAWR)
* The Electrical Safety, Quality and Continuity Regulations 2002 (ESQCR).

The HASWA 1974 is in place to cover all aspects of safety at work and can be viewed as the statutory document under which the other statutory documents which involve health and safety sit. The EAWR 1989 are specific to electrical installations used in the work place, although it is sensible for us to refer to them for all installations as this will ensure an electrically safe environment.

The HASWA and EAWR both have the word work in them and of course that reflects that they are intended for use in the work place. However, electricity is, or can be, a dangerous commodity wherever it is used. It should be remembered that a domestic installation is a place of work while any electrical work is being carried out; this means that the HASWA and EAWR are still relevant in most cases.

Non-compliance with statutory regulations is seen as a criminal offence, and for that reason non-compliance could result in a very large fine or even in some serious cases imprisonment, particularly where the non-compliance has resulted in an accident.

The ESQCR are intended more for electrical supplies but do have some effect on the daily activities of electricians, particularly with regards to the positioning of consumer units and areas where TNC systems are used. As an example, where an area is known to be susceptible to flooding all of the supply equipment and consumer units need to be sited above the expected flood level.

These statutory regulations not only apply to new installations, they also apply to existing installations which have been in use for a

Practical Guide to Inspection, Testing and Certification. 978-1-138-61332-4

very long time. There is no age limit on electrical installations; the requirement is that they are maintained in a safe condition and that they remain fit for use.

BS 7671 (non-statutory)

The most satisfactory way of ensuring conformance with statutory regulations is to follow the requirements of the relevant British Standard (BS). The BS relating to an electrical installation is known as BS 7671 and at the present time the industry is working to the 18th edition BS 7671 2018, which was published on 1 July 2018.

Within this set of standards Regulation 641.1 states 'every installation shall, during erection and on completion before being put into service be inspected and tested to verify, so far as reasonably practicable, that the requirements of the regulations have been met'. This regulation of course applies not only to new installations, it also applies to additions and alterations to existing installations.

The inspection and testing of new work is known as initial verification. As the regulation suggests, this initial verification commences at the same time as the installation work and continues throughout the duration of the job. The end result will be the issue of an *electrical installation certificate*, along with the required *schedule of test* results and a *schedule of inspections*, providing of course the work carried out fully meets the requirement of BS 7671.

As we have seen previously the EAWR 1989 are not only for new installations; if anything they are more relevant to existing installations.

BS 7671 Regulation 651.1 states that 'where required, periodic inspection and testing of every installation shall be carried out in accordance with Regulations 651.2 to 651.5 in order to determine, so far as is reasonably practicable, whether the installation is in a satisfactory condition for continued service'.

Although BS 7671 is a non-statutory document it has been referred to extensively in the Health and Safety Executive over a long period of time. Regulation 114 also clearly states that although BS 7671 is non-statutory it may be used in a court of law to claim compliance with a statutory requirement.

It has been my policy over the years to explain as clearly as possible to my students that although it is non-statutory, we can all do ourselves a big favour by pretending that it is statutory; this will ensure that we do the best job possible and that all safety requirements are met.

Building regulations

Part P of the building regulations is specific to electrical work carried out in domestic premises; however it is important to remember that whilst carrying out electrical work, all parts of the applicable building regulations must be complied with for all types of electrical installations.

Generally the other regulations which affect electrical installations are:

- Part A (Structure): this could be notches or holes in floor or roof joists, or the depth chases in walls.
- Part B (Fire safety): this would cover a range of things such as fire detection systems, fire and smoke alarms, fire resistance of floors, walls and ceilings.
- Part C (Site preparation and resistance to contaminants and water): any service penetrations must be resistant to rainwater and other contaminants such as radon. (Radon is a colourless and odourless radioactive gas which is formed by the radioactive decay of the small amounts of uranium which occur naturally in all soils and rock. It is thought that exposure to radon is the cause of around 1,000 cancer deaths a year.)
- Part E (Resistance to the passage of sound): where services pass through parts of the structure which are used to restrict the passage of sound, precautions must be taken to ensure that the structure is not compromised.
- Part F (Ventilation): where mechanical ventilation is installed, the ventilation rates must meet the current ventilation standards.
- Part L (Conservation of fuel and power): lighting must meet the requirements for energy efficiency.
- Part M (Access to and the use of buildings): this concerns the height of socket outlets and switches.

This is not a concise list and other building regulations may apply to special types of installations.

Domestic installations have been the subject of much discussion over the years, mainly due to the upsurge in DIY. We all know that it is reasonably easy to get something working; making sure it is safe is often far more difficult. To try and get some kind of control over this, the Building Regulation Part P was introduced and came into effect on 1 January 2005; it was then amended on 5 April 2006, and again on 6 April 2013. The purpose of this document is to ensure electrical safety in domestic electrical installations.

P1 sets out the requirements for the design and installation. It states that:

> *Reasonable provision shall be made in the design and installation of electrical installations in order to protect persons operating, maintaining or altering the installation from fire or injury.*

The requirements apply only to parts of the electrical installation which are intended to operate at low or extra low voltage and are:

- in or attached to a dwelling.
- in the common parts of a building serving one or more dwellings. This does not include the power supplies to lifts as they are governed by their own set of regulations.
- in a building which receives its supply of electricity from a source located within or shared with a dwelling.
- in a garden or in or on land associated with a building where the electricity is from a source located within or shared with a dwelling.

The requirements with Part P will be met where the installation:

- affords appropriate protection against mechanical and thermal damage.
- does not present electric shock and fire hazards to people.

Scope

The scope of Part P applies to electrical installations which are in a dwelling house or flat and to parts of the installation which are:

- outside the dwelling; this could include outside lights or socket outlets, photovoltaic systems, pond pumps or air conditioning units.
- supplies to outbuildings such as sheds, detached garages and domestic greenhouses. Only the electrical supply to the buildings are under the scope of Part P. Work carried out within this type of outbuilding is outside of the scope of Part P.
- common access areas in blocks of flats such as corridors and staircases.
- in shared amenities of blocks of flats such as kitchens, laundries and gymnasiums.
- in business premises which are connected to the same meter as a domestic dwelling; this could be a shop below a flat which shares a common meter.

Part P does not apply to business premises which have their own meters or lifts in blocks of flats. Part P does apply to a lift within a single dwelling.

Notifiable work

Part P requires that certain types of electrical work which is carried out in a domestic installation is notified to building control. Notifiable work includes:

- the installation of a single circuit or multiple circuits; this of course would include the rewiring of existing installations and completely new installations.
- the replacement of a consumer unit.
- any major change to a circuit; this could be changing the type or rating of a protective device or the installation of an RCD/RCBO.
- any addition or alteration to existing circuits in a special location.
- a room containing a swimming pool or sauna heater.

A special location would be a room containing a bath or a shower; however the special location would only vertically be to a height of 2.25m above the bath or shower tray and horizontally 600mm from the edge of a bath or shower tray.

In a wet room or a shower room which does not have a tray, the horizontal distance extends 1.2m from the centre of the shower head.

Inspecting and testing of notifiable work

There are three procedures which may be used to certify that notifiable work complies with the requirements set out in the building regulations.

The easiest method of notification is self-certification by a *registered competent person.* However, if the work has been carried out by a person who is not registered, it is acceptable for the installer to employ a third party registered person to certificate the work on their behalf. The other method is for the installation to be certified by a *local building control body.*

Self-certification by registered competent person

To become a registered installer, it is necessary to become a member of one of the recognised certification bodies which operate a domestic installer's scheme. This would require the person carrying out the work to prove competence in the type of work which is being carried out, and the ability to inspect, test and certificate the work which he/she has carried out. Competence is usually assessed by a site visit from an inspector employed by the chosen scheme provider.

Where the installer is registered as a domestic installer, he or she must complete an *electrical installation certificate*, along with a *schedule of test results* and a *schedule of inspections*. This can be one of the model forms from BS7671, or a certificate which has been developed specifically for Part P installations. The installer must then notify their scheme provider of the work which has been carried out.

This must be done within 30 days. The scheme provider will notify both the local authority and the customer that the correct certification has been provided. An annual fee is usually required by the scheme provider, while a small fee is also payable for each job registered; usually the fee includes insurance for the work carried out.

Certification by a registered third party

Where notifiable work has been carried out by a non-registered person, it is permissible for the installer to employ the services of a registered installer to carry out the certification for them.

Of course the installation must still meet the requirements of BS 7671 and the installation must be inspected and tested within five days of completion.

Where a third party certifier has been employed to carry out the inspecting and testing, the documents which must be completed are an *electrical installation condition report*, a *condition report inspection schedule* and a *schedule of test results*. The third party certifier must then notify their registration body who in turn will notify the customer and the building control body.

The building regulations state that when building work is complete the building should be no more unsatisfactory than it was before the building work was started. Therefore when altering or extending an electrical installation all new work must meet the current standards.

As a general rule, where an electrical installation requires an electrical installation certificate to be issued, the earthing and bonding arrangements for that circuit must meet the requirements of the latest edition of BS 7671. This will require that all of the earthing and bonding conductors are the correct size.

Where an electrical installation minor works certificate has been issued, there is no requirement to upgrade the earthing and bonding arrangements, providing of course that the installation is safe and that all of the required disconnection times can be met.

Non-notifiable work

Non-notifiable work would be additions and alterations to existing circuits which are not in a special location. This would also include repair, replacement and maintenance anywhere.

Installing fixed electrical equipment is within the scope of Part P even if it is connected using a standard 13 amp plug and socket. However, it is only notifiable if the work involves a new circuit.

The connection of electric cookers or other types of fixed equipment such as electric gates would not require notification unless the installation included a new circuit. Connecting to existing isolators is not notifiable.

Inspection and testing of non-notifiable work

Although the work may not be notifiable, it must still be certificated. The certificate used for minor work is called an electrical installation minor works certificate. Once completed the certificate must be given to the customer, with a copy being kept for your own records.

Always remember, certificating electrical work is just another part of the job which must be completed satisfactorily. It is not an EXTRA!!

All electrical work carried out in a dwelling must be thoroughly inspected and tested to verify that it is safe to be put into service, regardless of whether it is a domestic installation or not.

Building regulation compliance certificates or notices for notifiable work

This tells us that the completion certificates issued by the local authorities, etc. are not the same as the certificates that comply with BS 7671. The completion certificates do not only cover Part P, but also show compliance with all building regulations associated with the work which has been carried out.

Provision of information

Information should be provided for the installation to assist with the correct operation and maintenance. This information would comprise of:

• electrical installation certificates or reports describing the installation and giving details of work which has been carried out.

- permanent labels, for example on earth connection and bonds, and on items of electrical equipment such as consumer units and residual current devices.
- operating instructions of equipment belonging to the fixed installation and any logbooks which may be required.

For complex or unusually large installations detailed plans may be required.

Part P of the building regulations is for safety only; the functionality of the installation such as correct operation of electrically powered products (for example, ventilation and fire alarms) is covered by other parts of the building regulations.

Due to the introduction of Part P even people who are not in the electrical industry are becoming more and more aware that electrical installations need to be safe. Insurance companies and mortgage lenders are now frequently asking for certification as part of the house buying and selling process. The owners and occupiers of industrial and commercial properties are aware that the EAWR 1989 demand that they maintain a safe environment for people to work in, while most licensing authorities and local authorities are asking for electrical certification for most of the work with which they become involved.

All of these regulations are under the umbrella of the Health and Safety at Work Act 1974. This clearly puts the legal responsibility of health and safety on all persons concerned.

Summary

The term 'competent person' has been removed from the definitions section of amendment 3 of BS 7671. The requirement now is that persons carrying out electrical work must be a skilled person (electrically), a person who possesses, as appropriate to the electrical work being undertaken, adequate education, training or practical skills and is able to perceive risks and avoid hazards which electricity can create.

Many organisations provide what are known as Part P courses; however, it is not necessary to attend one of these in order to register as a domestic installer. While it may well be beneficial to an electrician who is a bit rusty to attend a refresher course just to ensure that they are aware of the requirements of Part P, it is not possible to become an electrician in 5 days! You will even see advertised courses lasting from 15 to 30 days; this is really just selling a dream, at the end of the period you will have spent a lot of hard earned cash and collected a lot of certificates. The one thing which you will not have is experience and that is the most important tool which you could possibly have in your box.

The buildings control authorities must be informed of any electrical work that is to be carried on a domestic electrical installation other than very minor work, although even this work must be certificated.

Building control can be informed (before commencing work) by the use of a building notice, and this will involve a fee.

If your work involves a lot of domestic electrical work, then by far the best route would be to join one of the certification bodies. This would allow you to self-certificate your own work. When you join one of these organisations, you must be able to show that your work is up to a satisfactory standard and that you can complete the correct paperwork (test certificates). Whichever organisation you choose to join, they will give you the correct advice on which training you require.

A qualification is fine, but being able to carry out electrical work safely is far better: for that reason high quality training is very important.

CHAPTER 2

Types of certification required for the inspecting and testing of electrical installations

Certification required for domestic installations (Part P)

The certification requirements for compliance with Part P are similar to the requirements for any other electrical installation.

It is a legal requirement to complete a Minor Electrical Installation Works Certificate (commonly called a 'Minor Works Certificate' or an 'Electrical Installation Certificate' for any electrical work being carried out on a domestic installation).

Minor electrical installation works certificate

This is a single document that must be issued when an alteration or addition is made to an existing circuit. A typical alteration that this certificate might be used for is the addition of a lighting point or socket outlet to an existing circuit. This certificate should be used for any installation regardless of whether it is domestic or not.

Part P domestic electrical installation certificate

Most registration bodies have specific domestic installation certificates which they like installers to use for work carried out in a dwelling. To achieve compliance with BS 7671, a certificate which contains all of the information shown on the certificates in BS 7671 Appendix 6 will be suitable; these can be purchased, downloaded or produced by the installer.

The completion of an electrical installation certificate is required when there has been

- a new installation
- an alteration
- an addition to an existing installation.

A *new installation* could be a completely new or rewired installation. An *alteration* could be a change of consumer unit or the installation of an RCD or new protective device. In reality, in most cases an alteration would be pretty much anything which resulted in a change of protection device without a change to the circuit conductors, although there will of course be the odd exception.

An *addition* would be where a new circuit or circuits are added to an existing installation.

In BS 7671 the electrical installation certificate is shown as a separate document. For the document to be valid it must be accompanied by a schedule of test results and a schedule of inspection. Many registration bodies prefer to produce a single certificate which contains all three of the required documents. These will be fully explained later in this book.

Periodic inspection, testing and reporting

There is no requirement in Part P for periodic inspection, testing and reporting. However, if the replacement of a consumer unit has been carried out, then the circuits which are reconnected should be inspected and tested to ensure that they are safe. This will, of course, require the following documentation:

- an electrical installation condition report
- a condition report inspection schedule
- a schedule of test results.

It is not a requirement of Part P that specific Part P certificates are used but you will find that many clients/customers prefer them.

The certificates produced by the IET (previously known as the IEE) are sufficient to comply with Part P and can be downloaded from the IET website (www.theiet.org) as described in the general certification section (Chapter 6).

Some documents contain a *schedule of items tested*, which can also be found on the IET website. Although it is not a requirement that this document is completed, it is often useful as a checklist.

Certification required for the inspecting and testing of installations other than domestic

(Further explanation is provided for these documents later in the book.)

All of these certificates are readily available from many sources. The basic forms can be downloaded from the IET website (www.theiet.org).

The NICEIC have forms which can be purchased by non-members and most instrument manufacturers produce their own forms, which are also available from electrical wholesalers. There are also available computerised programs which do the job really well, such as the Megger power suite, which is very easy to use and can also work with Bluetooth if you want to get really clever.

Minor electrical installation work certificate

This is a single sided A4 document (Figure 2.1) which should be used when minor alterations or additions are carried out on a circuit; this could be the addition of a socket outlet or perhaps moving a switch. It is possible to use an electrical installation certificate for this as well, although it would result in far more paper work.

Electrical installation certificate

This certificate must be issued for a completely new installation, an addition to an existing installation or an alteration to an existing installation. This would include any alterations to a circuit which would result in a change in the type of circuit protection provided. The changing of a consumer unit or the installation of an RCD would require this type of certificate to be completed.

The electrical installation certificate (Figure 2.2) must be accompanied by a schedule of inspections (Figure 2.3) and a schedule of test results (Figure 2.4). Without these two documents the electrical installation certificate is not valid.

This certificate is used to provide evidence of compliance with BS 7671; unless the work is completed and is to the standard required it must not be issued. An inspection and test which is carried out on a new installation to prove compliance with BS 7671 is known as an *initial verification.*

MINOR ELECTRICAL INSTALLATION WORKS CERTIFICATE

To be used only for minor electrical works which do not include the provision of a new circuit

---

Part 1 : Description of the minor work carried out

1. Details of the client... Date work completed................................

2. Installation address or location...

3. Description of the minor work carried out..

4. Details of any departures from BS 7671: 2018 for the circuit worked on. See regulations 120.3, 133.1.3 and 133.5. Where applicable a suitable risk assessment must be completed and attached to this document.

   Risk assessment attached ☐

5. Comments on the existing installation (including any defects observed. See regulation 644.1.2)

   ....

Part 2: Presence and suitability of the installation earthing and bonding arrangements (see regulation 132.16)

1. System earthing arrangement : TT ☐ TN-S ☐ TN-C-S ☐

2. Earth fault loop impedance (Z_{db}) at the distribution board supplying the final circuit.................Ω

3. Presence of adequate main protective conductors:

Earthing conductor ☐

Main protective bonding conductor(s) to: Water ☐ Gas ☐ Oil ☐ Structural steel ☐ Other......................... ☐

Part 3: Circuit details

DB reference No:............ DB location and type ....

Circuit No:........................ Circuit description..

Conductor sizes: Live conductors................mm^2 CPC...................mm^2

Circuit overcurrent protective device: BS(EN)................. Type...............Rating...................A

Part 4: Test results for the altered or extended circuit (where relevant and practicable)

Protective conductor continuity: R_2 .... Ω or $R_1 + R_2$..........................Ω

Continuity of ring final circuit conductors: L/L............Ω N/N..............Ω CPC/CPC..............Ω

Insulation resistance: Live - Live...............MΩ Live - Earth.............MΩ

Polarity correct: ☐ Maximum measured earth fault loop impedance: Z_s ....Ω

RCD operation: ☐ Rated residule operating current ($I_{\Delta n}$).....................mA

Disconnection time ....ms Test button operation is satisfactory ☐

Part 5: Declaration

I certify that the work covered by this certificate does not impair the safety of the existing installation and that the work has been designed, constructed, inspected and tested to the best of my knowledge and belief, at the time of my inspection complied with BS 7671 except as detailed in Part 1 of this document.

Name... Signature..

For and on behalf of... Position..

Address.. Date...

Figure 2.1 Minor electrical installation works certificate

ELECTRICAL INSTALLATION CERTIFICATE
(REQUIREMENTS FOR ELECTRICAL INSTALLATIONS – BS 7671 IET wiring regulations)

Details of the client

Installation address

Description and extent of the installation Description of the installation	New Installation ☐
Extent of the installation covered by this certificate:	Addition to an existing installation ☐
	Alteration to an Existing installation ☐

For Design

I/We being the person(s) responsible for the design of the electrical installation (as shown by the signatures below). The particulars of which are described above, having exercised reasonable care and skill when carrying out the design and also where the certificate applies to an addition or alteration, the safety of the existing installation is not compromised. I/We hereby certify that the design work for which I/we have been responsible is to the best of my belief and knowledge in accordance with BS 7671: 2018, amended to .... (date) except for any departures as detailed below.

Details of departures from BS 7671 (Regulations 120.3, 133.1.3 and 133.5)

Details of permitted exceptions (reg 411.3.3). Where applicable a suitable risk assessment must be completed and attached to this certificate

Risk assessment attached ☐

The extent of the liability is limited to the work described as the subject of this certificate

For the design of the installation:

Signature .... Date .... Name (in Block Letters)... Designer No 1

Signature .... Date .... Name (in Block Letters)... Designer No 2 where applicable

For Construction

I being the person(s) responsible for the construction of the electrical installation (as shown by the signatures below). The particulars of which are described above, having exercised reasonable care and skill when carrying out the construction. I hereby certify that the design work for which I have been responsible is to the best of my belief and knowledge in accordance with BS 7671: 2018, amended to .... (date) except for any departures as detailed below.

Details of departures from BS 7671 (Regulations 120.3 and 133.5)

The extent of the liability is limited to the work described as the subject of this certificate.

For **Construction** of the installation: Signature... Date ....

Name (In Block Letters)..For construction

For Inspection and Testing

I being the person(s) responsible for the inspecting and testing of the electrical installation (as shown by the signatures below). The particulars of which are described above, having exercised reasonable care and skill when carrying out the Inspecting and Testing. I hereby certify that the design work for which I have been responsible is to the best of my belief and knowledge in accordance with BS 7671: 2018, amended to .... (date) except for any departures as detailed below.

Details of departures from BS 7671 (Regulations 120.3 and 133.5)

The extent of the liability is limited to the work described as the subject of this certificate.

For **Inspection and Testing** of the installation: Signature... Date ....

Name (In Block Letters)..Inspector

Next Inspection

I/We being the designer(s) recommend that this installation is subjected to a periodic inspection and test after an interval of not more than ....years/months

---

Figure 2.2a Electrical installation certificate

Particulars of Signatories to the electrical installation certificate

Designer

Name: .... Company: ....

Address: ....

.... Postcode: .... Telephone No: ....

Constructor

Name: .... Company: ....

Address: ....

.... Postcode: .... Telephone No: ....

Inspector

Name: .... Company: ....

Address: ....

.... Postcode: .... Telephone No: ....

Supply Characteristics and Earthing Arrangements

Earthing arrangements	Number and type of live conductors		Nature of Supply Parameters	Supply Protective Device
TT ☐	AC ☐	DC ☐	Nominal Voltage, U /U$_o$ $^{(1)}$ V	BS (EN) ....
TN-S ☐	1-phase, 2 wire ☐	2-wire ☐	Nominal frequency, f$^{(1)}$ ....Hz	
TN-C-S ☐	2-phase, 3 wire ☐	3-wire ☐	Prospective fault current, I$_{pf}$$^{(2)}$ ....kA	Type..
TN-C ☐	3-phase, 3 wire ☐	other ☐	External loop impedance, Z$_e$$^{(2)}$.................Ω	
IT ☐	3-phase, 4 wire ☐		Note (1) by enquiry	Rated Current.....................................A
			(2) By measurement or enquiry	

Other sources of supply (As attached information)

Particulars of The Installation Referred to on the Certificate

Means of Earthing	Maximum Demand
Distributors facility ☐	Maximum Demand load..kVA/Amps (delete as necessary)
Installation earth electrode ☐	**Details of the installation earth electrode if installed**
	Type (e.g. Rods, tape, plate etc)...
	Location..
	Electrode resistance to earth......................Ω

Main Protective conductors

Earthing conductor	Material.................................. csa ....mm^2	Connection / continuity verified ☐
Main protective bonding conductors (to extraneous conductive parts)	Material..............................csa.............................mm^2	Connection / continuity verified ☐

To water installation pipes ☐	To gas installation pipes ☐	To oil installation pipes ☐	To structural steel ☐	To lighting protection ☐

To other ☐ ....

Main Switch / Fuse / Circuit Breaker / RCD

Location...	Current rating..A	**Where RCD is main switch**
....	Fuse / device rating...............................A	Rated residual operating current (I$_n$)...................mA
BS(EN)..	Voltage rating V	Rated time delay...ms
Number of poles..		Measured operating time ....ms

Comments On The Existing Installation (Where this is an addition or alteration see regulation 644.1.20)

....

....

....

....

....

Schedules

The attached schedules form part of this document and this circuit is only valid when they are attached to it

Number of schedules of inspection...Number of schedules of test results ....are attached

Figure 2.2b Continued.

Schedule of Inspections (for new installation work only) for Domestic and similar premises with a supply up to a 100 Amp supply

All items applicable to the installation work carried out must be inspected to confirm, where relevant, compliance with BS 7671. The list is not exhaustive and on some installations other items may require inspection.

A ✓ should be inserted where the inspection is satisfactory and N/A inserted where the area of inspection is not relevant to the work which has been carried out.

ITEM NUMBER	DESCRIPTION OF ITEM INSPECTED	Outcome ✓ or N/A
1.0	**External condition of intake equipment** (Visual inspection)	
1.1	Service Head	
1.2	Service Cable	
1.3	Earthing arrangements	
1.4	Meter tails	
1.5	Metering equipment	
1.6	Isolator if installed	
2.0	**PARALLEL OR SWITCHED ALTERNATIVE SOURCES OF SUPPLY**	
2.1	Adequate arrangements where a generating set operates as a switched alternative to the public supply (551.6)	
2.2	Adequate arrangements where a generating set operates in parallel with the public supply (551.7)	
3.0	**Automatic Disconnection of Supply**	
3.1	**Presence and adequacy of earthing and protective bonding arrangements:**	
	Distributors earthing arrangements (542.1.2.1 542.1.2.2)	
	Earthing conductor and connections, to include accessibility (542.3 543.3.2)	
	Installation earth electrode (if installed: 542.1.2.3)	
	Main protective bonding conductors and connections including accessibility (411.3.1.2 543.3.2 544.1)	
	Provision of electrical safety labels where required (514.13)	
	RCD(s) provided for fault protection (411.4.204 411.5.3) (not for additional protection)	
4.0	**Basic Protection**	
4.1	**Presence and adequacy of measures used to provide basic protection within the installation**	
	Insulation of live parts to comply with (416.1)	
	Barriers and enclosures have the correct IP rating (416.2)	
5.0	**Additional Protection**	
5.1	**Presence and effectiveness of additional protection methods**	
	RCD(s) with and operating current not exceeding 30mA (415.1 & Part 7) see section 8.1 of this schedule	
	Supplementary bonding (415.2 & part 7)	
6.0	**Other Methods of Protection**	
6.1	**Presence and effectiveness of methods used to provide both basic and fault protection**	
	SELV system, to include the source and associated circuits (414)	
	PELV system, to include the source and associated circuits (414)	
	Double or reinforced insulation such as class II equipment and associated circuits (412)	
	Electrical separation for one item of equipment such as a shaver socket (413)	
7.0	**Consumer units and distribution boards**	
7.1	Suitability of access and working space for items of electrical equipment and switchgear (132.12)	
7.2	Components are installed / fixed as manufacturer's instructions	
7.3	Presence of linked main switches where required (462.1.201)	
7.4	Isolators installed for each circuit or groups of circuits and individual items of equipment (462.2)	
7.5	Suitability of enclosures to comply with IP and fire ratings (416.2 421.1.6 421.1.201 and 526.5)	
7.6	Protection against mechanical damage to cables where they enter equipment (522.8.1 522.8.5 522.8.11)	
7.7	Confirmation that all conductors are correctly located and are secure and tight (526.1)	
7.8	Avoidance of heating effects where cables enter ferromagnetic enclosures eg Steel consumer units (521.5)	

Figure 2.3a Schedule of inspections

ITEM NUMBER	DESCRIPTION OF ITEM INSPECTED	Outcome ✓ or N/A
	Consumer units and distribution boards continued	
7.9	Selection of the correct type and rating of protective devices for overcurrent and fault protection (411.3.2 411.4 411.5 411.6 sections 432 433 537.3.1.1)	
7.10	**Presence of appropriate circuit charts and other notices**	
	Provision of circuit charts /schedules or other forms of information (514.9)	
	Isolation warning notices where live parts cannot be isolated by a single device (514.11)	
	Periodic inspection and testing notice (514.12.1)	
	Presence of six monthly test notice for RCDs where required (514.12.2)	
	Presence of a six monthly test notice for any AFDDs which have been installed	
	Warning notice of non standard colours of conductors has been fitted (514.4)	
	Correct labelling of switchgear, isolators and protective devices (514.1.1 514.8)	
8.0	**Circuits**	
8.1	Adequacy of conductors for current carrying capacity with regard to the type and nature of the installation (section 523)	
8.2	Cable installation methods suitable for the location and any external influences (section 5.22)	
8.3	Segregation/separation of any Band I (ELV) and Band II (LV) circuits and electrical and non-electrical services (528)	
8.4	Cables correctly erected and supported throughout with protection against abrasion (sections 521 528)	
8.5	Provision of fire barriers and sealing arrangements where required (527.2)	
8.6	Non sheathed cables enclosed throughout in conduit ducting or trunking (521.10.1 526.8)	
8.7	Cables concealed in walls and partitions, under floors, above ceilings, or in voids protected against damage. (522.6.201 to 204)	
8.8	Conductors correctly identified by colouring, lettering or numbering (section 514)	
8.9	Presence, suitability and correct termination of protective conductors (411.3.1.1 543.1)	
8.10	Cables and conductors correctly connected and with no undue mechanical strain (Section526)	
8.11	No basic insulation of a conductor visible outside of an enclosure (526.8)	
8.12	Single-pole devices for switching or protection installed in line conductors only (132.14.1 530.3.3 643.6)	
8.13	Accessories not damaged, correctly connected, securely fixed and suitable for external influences (134.11 512.2 section 526)	
8.14	Provision of additional protection requirements by RCD not exceeding 30mA	
	Socket outlets rated at 32A or less, unless exempt (411.3.4)	
	Supplies for mobile equipment with a current rating not exceeding 32A for use outdoors (411.3.3)	
	Cables concealed in walls at a depth of less than 50mm (522.6.202 522.6.203)	
	Cables concealed in wall containing metal parts regardless of depth (522.6.202 522.6.203)	
	Final circuits supplying luminaires within domestic (household) properties (411.3.4)	
8.15	**Presence of the correct devices for isolation and switching correctly located**	
	Means of switching off for mechanical maintenance (section 464 537.3.2)	
	Emergency switching (465.1 537.3.3)	
	Functional switching for control of parts of the installation and current using equipment (463.1 537.3.1)	
	Fire fighters switches (537.4)	
9.0	**Current using equipment**	
9.1	Equipment not damaged, securely fixed and suitable for external influences (134.1.1 416.2 512.2)	
9.2	Provision of overload and/or undervoltage protection where required (section 445 5520)	
9.3	Installed to minimise the build-up of heat and spread of fire (421.1.4 559.4.1)	
9.4	Adequacy of working space, accessibility of equipment (132.12 513.1)	
10.0	**Locations containing a bath or a shower (section 701)**	
10.1	30 mA RCD protection for all LV circuits, equipment suitable for zones. Supplementary bonding installed where required	
11.0	**Any other part 7 special installations or locations**	
11.1	List all other special locations or installations present and record on a separate page any particular inspections carried out	

Inspected by ....

Name in capitals... Signature..Date..................................

Figure 2.3b Continued.

GENERIC SCHEDULE OF TEST RESULTS

DB Reference no..........................

Details of circuits and/or installed equipment which may be vulnerable to damage when testing.................

Details of test instruments used (state serial or asset numbers)

Location...................................
Z_s at DB (Ω).............................
I_pf at DB (kA)...........................
Correct supply polarity confirmed ☐
Phase sequence confirmed (where appropriate)

Continuity.................................
Insulation resistance....................
Earth fault loop impedance............
RCD...
Earth electrode resistance.............

Tested by:
Name in capitals.........................
Signature..................Date..........

Test Results

Circuit number	Circuit description	BS (EN)	Type	Rating (A)	Breaking capacity (kA)	RCD I_Δn mA	Maximum permitted Z_s	Ring final circuit continuity (Ω) — r1 (Line)	rn (neutral)	r2 (cpc)	Continuity (Ω) (R_1 + R_2) or R_2 — (R_1 + R_2)	R2	Insulation resistance test — Volts	Insulation resistance (MΩ) — Live - Live	Live - Earth	Polarity	Z_s (Ω) Maximum measured	RCD — Disconnection Time (ms)	RCD test button operation	AFDD — Manual AFDD test button operation	Remarks Use separate sheet if required

Where the tabulated values of maximum permitted earth fault loop impedance is taken from a source other than chapter 41 of BS 7671, state the source of the information in the remarks column of this schedule

Figure 2.4 Schedule of test results

CHAPTER 3

Initial verification inspection

Introduction

The documentation which should be completed is the electrical installation certificate; this must always be accompanied by a schedule of test results and a schedule of inspection.

The purpose of this inspection is to verify that the installed equipment complies with BS or BS EN standards; that it is correctly selected and erected to comply with BS 7671; and that it is not visibly damaged or defective so as to impair safety (Regulation 642.2).

When a new installation has been completed, it must be inspected and tested to ensure that it is safe to use. This process is known as the initial verification (Regulation 641.1). For safety reasons, the inspection process must precede testing and the part of the installation must be complete.

Regulation 641.1 clearly tells us that the inspecting and testing process must be ongoing from the moment the electrical installation commences. In other words, if you are going to be responsible for completing the required certification, but are not the person carrying out the work, you must visually inspect any parts of the installation which will eventually be covered up while they are still exposed. This of course will require that you visit the installation at regular intervals as the work progresses.

Remember! If you are signing the certificate you are the person who will probably be held responsible if at a later date defects are found which can be traced back to the original installation. Items like undersized cables and high Z_s values caused by long circuits will have originated from the original installation and will always be traceable.

Practical Guide to Inspection, Testing and Certification. 978-1-138-61332-4

For this reason, by the time the installation is completed and ready for certification, a great deal of the installation must have already been visually inspected.

As an initial verification is ongoing from the commencement of the installation and much of the required inspecting and testing will be carried out during the installation, it is important that the whole range of inspection and tests are carried out on all circuits and outlets. Clearly it would not be sensible to complete the installation and then start dismantling it to check things like tight connections, fitting of earth sleeving and identification of conductors, etc.

There are many types of electrical installations and the requirements for them will vary from job to job. Every installation should be treated individually. Where relevant, the following items should be inspected to ensure that they comply with BS 7671; as previously mentioned some items will require inspecting before they are hidden in building voids and parts of the building structure. This would need to be carried out during erection.

The following are examples of the items requiring inspection.

Electrical intake

- Condition of the service cable.
- Service head – is it secure or damaged? On older installations the service head would have been filled with pitch; the sight of this having seeped out of the bottom of the head will indicate a possible overload.
- Supply earthing arrangements – is the conductor the correct size? Has the clamp corroded? Is the clamp tight?
- Are the meter tails the correct size? Are they double insulated?
- Is the metering equipment fixed securely? Has it been tampered with?
- On installations where the supply has an isolator, is the isolator the correct rating? Is it secure?

Alternative supply sources

These could be photovoltaic installations, wind turbines or standby generators.

- Are there suitable arrangements to ensure that the alternative system can operate safely?

- Does the generating system have a dedicated earthing system which is independent of the public supply?
- Has the system been correctly connected in parallel with the supply?
- Is the system compatible with the installation?
- Does the generation system have a means of isolation from the public supply?
- Where the generation system is a photovoltaic array or a wind turbine, has a means of automatic disconnection from supply been provided when there is a loss of public supply? This is usually the inverter.

Automatic disconnection of supply

Automatic disconnection of supply is where the installation is set up so that in the event of a fault, the supply is interrupted before the touch voltages reach a dangerous level. If the fault is between two live conductors it is referred to as a short circuit, and if the fault is between live conductors and earth it is referred to as an earth fault.

In both cases disconnection can be achieved by the operation of a fuse or a circuit breaker; in the case of an earth fault, the earth fault loop path must be low enough to allow enough current to flow to either melt the fuse or trip the circuit breaker.

In situations where a low earth fault path resistance cannot be achieved, an RCD will need to be installed. It must be remembered however that an RCD will only offer protection against earth faults and will not offer protection against short circuit.

On electrical documents which require the method of fault protection to be listed, it is usual to abbreviate automatic disconnection of supply to ADS.

- Are the required earthing and protective bonding arrangements in place?
- Are the distributor's earthing arrangements suitable or does the installation require the use of an earth electrode?
- Is the earthing conductor connected/terminated correctly and is it the correct size?
- Are the main protective bonding conductors in place, correct size, terminated correctly using the correct methods with the correct labelling?
- Are the earthing conductor and protective conductor terminations accessible?

Basic protection

Basic protection is exactly what it sounds like, it is the most basic method of preventing anything from touching electrical parts which are intended to be live, or if the live parts can be touched that the voltage is so low that it will not be harmful. Clearly all installations must have some form of basic insulation.

- Are all live parts insulated correctly with insulation which can only be removed by destruction?
- Do all of the barriers and enclosures meet the correct IP ratings? This would be a minimum of IPXXB or IP2X for all surfaces other than a horizontal top surface of an enclosure which is readily accessible; in these circumstances the level of protection would need to be a minimum of IPXXD or IP4X.
- Are obstacles, where used, suitable to provide the required level of protection?
- Does placing out of reach, where used, provide a suitable level of protection? Are bare overhead lines the correct height for example?

Fault protection

Fault protection is a combination of the use of protective devices and protective bonding; it is intended to prevent damage and electric shock in the event of an electrical fault. In situations where there is a short circuit, or a low resistance fault to earth, the installation design should be such that a protective device will operate and interrupt the supply. In some instances the magnitude of the fault may not be enough to operate a protective device within the required time. This could be where the fault is not of a sufficiently low resistance to allow enough current to flow to operate the short circuit part of a circuit breaker or fuse. Where the fault is an earth fault, protective bonding is required; this allows the touch voltage between exposed and extraneous conductive parts to rise to the same potential. Protective bonding is also important in situations where for some reason an RCD fails to operate, or perhaps a protective device fails to open due to a mechanical fault.

Where RCDs are used for fault protection they should not be confused with those which are used for additional protection; although the devices are the same the reason for their use is entirely different. This difference will be explained further later in the book. As well as ADS, the following methods can be used as fault protection:

- non-conducting location (rarely used);
- earth free local equipotential bonding (rarely used; in laboratories perhaps or medical areas);
- electrical separation (most common is the use of shaver sockets). Does the transformer meet the correct standard? This type of protection may also be used in electrical repair workshops for safety reasons.

Basic and fault protection

These are methods which are acceptable for basic and fault protection and are said to be provided where:

(i)   The nominal voltage cannot exceed the upper limit of band I; this is normally not exceeding 50 volts a.c. or 120 volts d.c.

(ii) The supply is either from:
- a safety isolating transformer which meets the BS requirements
- or a source providing a degree of safety equivalent to an isolating transformer or an electromechanical source such as a battery.

(iii) There is an electronic device which complies with the required standard, and where provisions have been taken to ensure that an internal fault will not cause the outgoing terminals to a voltage higher than SELV or PELV.
- Where SELV and PELV is used, there needs to be confirmation that the requirements are satisfied.
- Where double and reinforced insulation is used, there needs to be confirmation that all of the requirements are satisfied.

Additional protection

This should not be confused with fault protection. Additional protection is exactly what it sounds like; it is a form of protection which is put in place to supplement the other types of protection which have been provided. This is usually required where there is an increased risk of danger; examples of this would be the use of socket outlets in a domestic environment or circuits supplying equipment in rooms containing a bath or a shower.

- RCDs used for additional protection must not exceed 30 mA and must go open circuit within 40 ms when there is a current of $5 \times$ the rated operating current ($I_{\Delta n}$).
- Supplementary bonding can also be used for additional protection and must be correctly installed.

Initial verification testing

During the initial verification, each circuit must be tested. This will require the use of the correct type of testing equipment which is detailed later in this book.

For safety reasons, during initial verification it is important that the testing procedure is carried out in the correct sequence as stated in Guidance Note 3 of BS 7671. This will also reduce the risk of having to go back and repeat tests which have already been carried out should a fault be found.

Sequence of tests

The sequence of tests is as follows:

- Continuity of bonding conductors and circuit protective conductors
- Continuity of ring final circuit conductors
- Insulation resistance
- Site applied insulation
- Protection by separation of circuits
- Protection by barriers and enclosures
- Insulation of non-conducting floors
- Dead polarity of each circuit
- Live polarity of supply
- Earth electrode resistance (Z_e)
- Earth fault loop impedance (Z_e)(Z_s)
- Prospective fault current (I_{pf})
- Additional protection
- Phase sequence
- Functional testing
- Verification of voltage drop.

Periodic inspection

A periodic inspection is carried out to ensure that an installation is safe and has not deteriorated over a period of time.

A periodic inspection would be carried out for many reasons. Examples are:

- the recommended due date
- change of occupancy
- change of use
- change of ownership
- insurance purposes

- mortgage requirement
- before additions or alterations
- after damage
- client request.

The approach to this type of inspection is very different from that for an initial verification. It is extremely important that where a periodic inspection is to be carried out on an electrical installation the original *electrical installation certificate*, or past *periodic inspection/condition reports*, along with the *schedules of test results* and the *schedules of inspection*, are available.

If this required documentation is not available, then the inspection and testing cannot proceed until a survey of the installation has been carried out and fuse charts along with any other documents that the inspector requires are prepared.

Where previous documentation is not available for either the whole installation, or even perhaps individual circuits which have been installed and not certificated, it is important that a full survey is carried out on the part of the installation for which there is no information. Where there is a record of Z_e then only the individual circuits will need to be isolated; however if there is no record of Z_e then the whole installation will have to be isolated to enable a Z_e value to be measured.

Each circuit must also have a recorded value of $R_1 + R_2$. Where these values are not available from original certificates, or previous periodic inspection and test reports, the isolation and dismantling of each circuit will be required.

In most situations where a periodic inspection is required, the installation will have been used, the building is often occupied and still in use. It may possibly have had additions and alterations made to it, and the type of use or even the environment could have changed from that which the installation was originally designed for.

Before commencing work the extent and limitation of the inspection must be agreed with the person ordering the work. The *extent* is the amount of the installation which is to be inspected; this decision will require that the person carrying out the inspection has experience in the type of installation which is to be inspected. A minimum of 10 per cent of the installation should be inspected; this could increase, depending on any defects found.

In many cases, it may not be possible to isolate the complete installation; this is often the case where an inspection is being carried out on larger industrial/commercial installations or perhaps public buildings. This does not really present a problem providing that the original or latest inspection and test documentation is available; this would also be noted as a limitation.

In some installations it will not be possible even to isolate some circuits due to the disruption which it could cause, or the client may request that a circuit is not isolated for a certain reason; this would be recorded as a **limitation**. Other limitations could be:

- areas not to be entered, these could be meeting rooms or food processing areas
- circuits not to be inspected
- times when circuits can be isolated
- times when areas could be accessed
- certain tests not to be carried out on circuits, this could be insulation resistance tests on circuits which may have vulnerable equipment connected.

Clearly this list is not exhaustive and the limitation area of the condition report should be used to record anything which is not going to be inspected.

Unlike an initial verification, the inspection should not be intrusive. Although covers will need to be removed in certain areas, it is not usually necessary to remove all accessories or carry out the full range of tests on every circuit. This will depend on what the inspector discovers as the inspection is carried out. Regulation 651.2 of BS 7671 makes it quite clear that a 'detailed examination of the installation shall be carried out without dismantling or with partial dismantling as required'. All too often electrical installations are damaged due to parts of the installation being taken apart when really there is no real need to.

This is where experience is very important, particularly on larger installations.

Visual inspection

When carrying out a visual inspection on an electrical installation there are basic areas of inspection which should be addressed. In general terms we are inspecting the installation with regards to:

- *Safety*
- *Age*
- *Deterioration*
- *Corrosion*
- *Overload*
- *Wear* and tear.

An easy way to remember this is to use the acronym *SADCOW.*

Suitability and external influence should also be included along with a close look at any alterations. At this point it is a good idea to get from the client any documentation that is relevant to the installation. These documents could include:

- plans
- drawings
- previous test results and certification
- fuse charts.

You should also make it clear that you will require access to all parts of the building and that the electricity will need to be turned off to allow you to carry out some of the tests. Of course it will not always to be possible for the supply to be turned off, and occasionally there will be some areas of the building which it may not be possible to enter for various reasons. In these cases they must be recorded as a limitation, as it is just as important to document anything which is not done as it is to document the work which is.

A visual inspection of any installation is as important as any testing which may need to be carried out. Remember it is called a periodic inspection – there is no mention of test. If you are not familiar with the building it is always a good idea to have a walk around before you start the serious work; during this time you will be able to identify any areas which may require particular consideration.

The first requirement of the visual inspection is to check that the system is safe to carry out a detailed inspection and also safe to carry out any tests which may be deemed necessary. Generally, a good place to start would be the supply intake; this will give you a reasonable indication of the age, type and size of the installation.

The documents which need to be completed for a condition report are in Appendix 5 of BS 7671, and Amendment 3 provides a form which is very comprehensive and is intended to be used as a checklist.

Things to look for at the supply intake *before* removal of any covers would be:

- The type of the supply system – is it TT, TNS or TNCS?
- Is it old or modern?
- Are the conductors imperial or metric?
- What type of protection is there for the final circuits?
- Is the consumer unit or distribution board labelled correctly?
- Is the earthing conductor in place?
- Is the earthing conductor the correct size?
- Is the earthing conductor green or green and yellow?
- Are all of the circuits in one consumer unit or are there two or three units which need combining?
- Is there any evidence of the main protective bonding? Remember it must start at the main earthing terminal.

ELECTRICAL INSTALLATION CONDITION REPORT

Section A. Details of the person ordering the report
Name ....
Address...
....

Section B. Reason for this report being produced ....
....
Date on which the inspection was carried out...

Section C. Details of the installation which is the subject of this report.
Occupier...
Address...
....

Description of premises
Industrial ☐　Commercial ☐　Domestic ☐　Other (description of premises)..................................
Estimated age of wiring system................Years
Any evidence of additions or alterations　Yes ☐　No ☐　Non apparent ☐　If yes estimate of age ....Years
Are installation records available (reg 651.1) Yes ☐　No ☐　Last inspection date..................................

Section D. Extent and limitations of inspection and testing carried out
Extent of the electrical installation covered by this report..
....
....
Agreed limitations and reasons (Reg 653.2)...
....
Agreed with...
Operational limitations with reasons ....

The inspecting and testing detailed in this report and attached schedules have been carried out in accordance with BS 7671 : 2018 as amended to ....
Cables concealed within conduits and trunking. Inaccessible roof voids, under floors and other building voids have not been inspected unless specifically agreed with all parties involved.

Section E. Summary of the condition of the installation
General condition of the installation in terms of electrical safety...
....
....
General assessment of the installation in terms of continued use　Delete as appropriate　SATISFACTORY　/　UNSATISFACTORY
An unsatisfactory assessment indicates that a part or parts of the installation are dangerous (code C1) and/or parts of the installation are potentially dangerous (code C2) and have been identified.

Section F. Recommendations
Where the overall assessment of the suitability for continued use is stated as UNSATISFACTORY, I/We recommend that any C1 (Dangerous) or C2 (potentially dangerous) observations are acted upon as a matter of urgency.
Any items marked as further investigation required (code F1) should be investigated as soon as reasonably possible.
Any improvement recommendations (Code C3) should be considered.

Subject to the required remedial work being carried out I/We recommend that this installation is inspected and tested no later than ....Date

Section G. Declaration
I/We being the persons responsible for the inspection and testing of the electrical installation, particulars of which are described above, having exercised reasonable skill and care when carrying out the inspection and testing, hereby declare that the information in this report, including the observations and the attached schedules provide an accurate assessment of the condition of the installation taking into account the stated extent and limitations listed in section D of this report.

Inspected and tested by:	**Report authorised for issue by:**
Name (Capitals)..	Name (Capitals)..
Signature...	Signature...
For and on behalf of..	For and on behalf of..
Position..	Position..
Address..	Address..
....	....
Date..	Date..

Section H : Schedule(s)
.... Schedule (s) of inspection and ....Schedules of test results are attached

The attached schedules form part of this document and the report is only valid when they are attached to it

Figure 3.1a Copy of condition report inspection

Section I. Supply Characteristics and Earthing Arrangements

Earthing arrangements	Number and type of live conductors		Nature of Supply Parameters	Supply Protective Device
TT ☐ TN-S ☐ TN-C-S ☐ TN-C ☐ IT ☐	AC ☐ 1-phase,2 wire ☐ 2-phase, 3 wire ☐ 3-phase,3 wire ☐ 3-phase, 4 wire ☐	DC ☐ 2-wire ☐ 3-wire ☐ other ☐	Nominal Voltage, U/U_0 [(1)] ....V Nominal frequency, f [(1)] ....Hz **Prospective fault current, I_{pf} [(2)] ....kA** External loop impedance, Z_s [(2)].................Ω Note (1) by enquiry (2) By measurement or enquiry	BS (EN) .... Type... Rated Current....................................A

Other sources of supply (As attached information)

Section J: Particulars of The Installation Referred to on the report

Means of Earthing	**Maximum Demand**
Distributors facility ☐	Maximum Demand load..kVA/Amps (delete as necessary)
Installation earth electrode ☐	**Details of the installation earth electrode if installed** Type (e.g. Rods, tape, plate etc).. Location... Electrode resistance to earth....................Ω

Main Protective conductors

Earthing conductor	Material.. csa ....mm^2	Connection / continuity verified ☐
Main protective bonding conductors (to extraneous conductive parts)	Material.........................csa...................................mm^2	Connection / continuity verified ☐

To water installation pipes ☐	To gas installation pipes ☐	To oil installation pipes ☐	To structural steel ☐	To lighting protection ☐

To other ☐ ....

Main Switch / Fuse / Circuit Breaker / RCD

Location... .... BS(EN).. Number of poles...................................	Current rating...A Fuse / device rating...............................A Voltage rating V	**Where RCD is main switch** Rated residual operating current (I_n)....................mA Rated time delay..ms Measured operating time ....ms

Section K: Observations

Referring to the attached schedules of inspection and test results, and subject to the limitations specified in section D

No remedial action required ☐   The following observations are made ☐

Observation(s) Include schedule reference	Classification Code
....	....

One of the following codes, as appropriate, has been allocated to each of the observations made above to indicate to the person responsible for the installation the degree of urgency for remedial action.

C1 – Danger present, immediate remedial action required

C2 – Potentially dangerous – urgent remedial action required

C3 – Improvement recommended

F1 – Further investigation required without delay

Figure 3.1b Continued.

CONDITION REPORT INSPECTION SCHEDULE FOR DOMESTIC AND SIMILAR PREMISES UP TO 100 A SUPPLY

OUTCOMES	Acceptable condition	✓	Unacceptable condition	State C1 or C2	Improvement recommended	State C3	Further investigation	F1	Not Verified	N/V	Limitation	LIM	Not Applicable	N/A
ITEM NO	**DESCRIPTION**								**OUTCOME** Use codes shown above and provide additional comments where required. Any item coded C1, C2, C3 and F1 must be recorded in section K of the condition report					

ITEM NO	DESCRIPTION	
1.0	External condition of intake equipment (visual check only)	
1.1	Service cable	
1.2	Service head	
1.3	Earthing arrangement	
1.4	Meter tails	
1.5	Metering equipment	
1.6	Isolator (where present)	

2.0	**Presence of suitable arrangements for other sources such as microgeneration (551.6, 551.7)**	

3.0	Earthing / Bonding Arrangements (411.3 Chap 54)	
3.1	Presence and condition of distributors earthing arrangement (542.1.2.1; 542.1.2.2)	
3.2	Presence and condition of the earth electrode if applicable (542.1.2.3)	
3.3	Presence of earthing and bonding labels in the correct locations (514.13.1)	
3.4	Confirmation of the earthing conductor size (542.3; 543.1.1)	
3.5	Accessibility and condition of the earthing conductor at the MET (543.3.2)	
3.6	Confirmation of any main protective bonding conductor sizes (544.1)	
3.7	Condition and accessibility of main protective conductor connection (543.3.2; 544.1.2)	
3.8	Accessibility and condition of other protective bonding connections (543.3.1 543.3.2)	

4.0	**Consumer unit (s) / Distribution boards**	
4.1	Adequacy of working space/access to consumer/distribution boards (132.12 : 513.1)	
4.2	Security of fixing (134.1.1)	
4.3	Condition of enclosures in terms of IP ratings etc (416.2)	
4.4	Condition of enclosures in terms of fire rating etc (421.1.201; 526.5)	
4.5	Enclosure not damaged so as to impair safety (651.2)	
4.6	Presence of main linked switch (462.1.201)	
4.7	Operation of main switch (functional check) (643.10)	
4.8	Manual operation of circuit breakers and RCDs to prove disconnection (643.10)	
4.9	Correct identification of circuits and protective devices (514.8.1; 514.9.1)	
4.10	Presence of RCD six-monthly test notice at/near consumer unit/distribution board (514.12.2)	
4.11	Presence of non-standard (mixed) cable colour warning notice at/near consumer unit/distribution board (514.14)	
4.12	Presence of alternative supply warning notice at/near consumer unit/distribution board (514.15)	
4.13	Presence of other required labelling (specify) (section 514)	
4.14	Compatibility of protective devices, bases and other components; correct type and rating (no signs of acceptable thermal damage, arcing and or overheating) (411.3.2; 411.4; 411.5; 411.6; sections 432,433)	
4.15	Single-pole switching or protective devices in the line conductor only (132.14.1; 530.3.3)	
4.16	Protection against mechanical damage where cables enter enclosures (132.14.1; 522.8.1; 522.8.5; 522.8.11)	
4.17	Protection against electromagnetic effects where cables enter ferrous enclosures (521.5.1)	
4.18	RCDs provided for fault protection- including RCBOs (411.3.3; 415.1)	
4.19	RCDs provided for additional protection including RCBOs (411.3.3; 415.1)	
4.20	Confirmation of information that SPD is functional (651.4)	
4.21	Confirmation that all conductor connections including connections to busbars, are correctly terminated and are tight and secure (526.1)	
4.22	Adequate arrangements where a generating set operates as a switched alternative to the public supply (551.6)	
4.23	Adequate arrangements where a generating set operates in parallel with a public supply (551.7)	

Figure 3.1c Continued.

OUTCOMES	Acceptable condition	✓	Unacceptable condition	State C1 or C2	Improvement recommended	State C3	Further investigation	F1	Not Verified	N/V	Limitation	LIM	Not Applicable	N/A
ITEM NO	DESCRIPTION								**OUTCOME** Use codes shown above and provide additional comments where required. Any item coded C1, C2, C3 and F1 must be recorded in section K of the condition report					

5.0	**Final Circuits**	
5.1	Identification of conductors (514.3.1)	
5.2	Cables correctly supported throughout their run (521.10.202; 522.8.5)	
5.3	Condition of insulation of live parts (416.1)	
5.4	Non-sheathed cables protected by enclosures in conduit, trunking or ducting (521.10.1)	
	• To include the integrity of the trunking and conduit systems	
5.5	Adequacy of cables for current carrying capacity with regard to the type and nature of the installation (section 523)	
5.6	Coordination between conductors and overload protective devices (433.1; 533.2.1)	
5.7	Adequacy of protective devices: type and rated current for fault protection (411.3)	
5.8	Presence and adequacy of circuit protective conductors (411.3.1; Section 543)	
5.9	Wiring system(s) appropriate for the type and nature of the installation and external influences (section 522)	
5.10	Concealed cables installed in prescribed zones (see section D, Extent and limitations) (5226.204)	
5.11	Cables concealed under floors, above ceilings or in walls/partitions, adequately protected against damage (see section D, Extent and limitations) (5226.204)	
5.12	Provision of additional requirements for protection by RCD not exceeding 30mA	
	• For all socket outlets of rating 32A or less, unless an exception is permitted (411.3.3)	
	• Supply of mobile equipment not exceeding 32A rating for use outdoors (411.3.3)	
	• For cables installed in walls at depth of less than 50 mm (522.6.202; 522.6.203)	
	• For cables concealed in walls/partitions containing metal parts regardless of depth (522.6.203)	
	• Final circuits supplying luminaires within domestic (household) premises (411.3.4)	
5.13	Provision of fire barriers, sealing arrangements and protection against thermal effects (section 527)	
5.14	Band II cables segregated / separated from Band I cables (528.1)	
5.15	Cables segregated / separated from communications cables (528.2)	
5.16	Cables segregated / separated from non-electrical services (528.3)	
5.17	Termination of cables at enclosures- indicate the extent of sampling in section D of the report (section 526)	
	• Connection soundly made and under no strain (526.6)	
	• No basic insulation of a conductor visible outside of an enclosure (526.8)	
	• Connection of live conductors adequately enclosed (526.5)	
	• Adequately connected at the point of entry into an enclosure(bushes, glands etc) (522.8.5)	
5.18	Condition of accessories including socket-outlets, switches and joint boxes (651.2(v))	
5.19	Suitability of accessories for external influences (512.2)	
5.20	Adequacy of working space/ accessibility of equipment (132.12; 513.1)	
5.21	Single-pole switching or protective devices in the line conductor only (132.14.1; 530.3.3)	

6.0	**Location(s) containing a bath or a shower**	
6.1	Additional protection for all low voltage circuits by an RCD not exceeding 30mA (701.411.3.3)	
6.2	Where used as a protection measure, the requirements of SELV and PELV are met (701.414.4.5)	
6.3	Shaver sockets comply with BS EN 61558-2-5 formerly BS 3535 (701.512.3)	
6.4	Presence of supplementary bonding conductors, unless not required by BS 7671:2018 (701.415.2)	
6.5	Low voltage socket-outlets sited at least 3 m from zone 1 (701.512.3)	
6.6	Suitability of equipment for external influences for the installed location in terms of IP rating (701.512.2)	
6.7	Suitability of accessories and control gear etc., for a particular zone (701.512.3)	
6.8	Suitability of current using equipment for the particular position within the location (701.55)	

7.0	**Other part 7 special installations or locations**	
7.1	List all other special locations or installations present, if any (record separately the results of any inspections applied)	

Inspected by;

Name (Capitals)..   **Signature**..   **Date**..

Figure 3.1d Continued.

- What size is the main protective bonding? Is it large enough?
- Is there a residual current device? If so, does it have a label attached?
- Does the installation meet the requirements with regards to RCD protection? In most installations a split board is required as a minimum.
- If there is an RCD, is it current operated or voltage operated?
- Are there any RCBOs?
- Do the enclosures meet the required IP codes (Regulation 416.2)?
- If alterations have been carried out since January 2005, has a warning notice been fitted on or near the distribution board to indicate that harmonised colour cables have been used (Regulation 514.14.1)?
- Where alterations have been carried out, is there documentation available for them, along with test results?
- What is the type and rating is the supply fuse? Is it large enough for the required load?
- Are the meter tails large enough?
- Are the seals broken on the supply equipment? If they are, it could indicate that the system has been tampered with since it was first installed and perhaps a closer investigation is required.
- Have any alterations or additions been made?
- Is it possible to remove covers without having to dismantle parts of the building such as cupboards?
- Has the installation got any other supply systems such as photovoltaic systems? If so, are the isolation points identified correctly with dual supply labels?

This list is not exhaustive and each installation will have its own requirements.

When the visual inspection of the supply intake area is complete, that is a good time to look around the rest of the building to make sure that there are not any very obvious faults. All of this should be carried out before the removal of any covers.

Things to look for during your walk around include:

- Are accessories fixed to the wall properly? Are they missing or damaged?
- Are the accessories old with wooden back plates?
- Are the socket outlets round pin or square? Is there a mixture of both? If there is, it is a good indication that the installation has been altered, possibly by the householder or one of his/her friends.
- Have cables been installed where there is a possibility of them being damaged? Clipped up the skirting with no protection is a common one.

- Have cables been clipped correctly?
- Have enclosures been fixed securely?
- Have ceiling pendants got perished flexes? Particular attention should be given to old braided and rubber type flexes.
- Are all parts of the installation which is installed out doors suitable for the environment (IP ratings)?
- Are all earthing clamps compliant with BS 951 and correctly labelled?
- If extraneous conductive parts such as water and gas are bonded using the same protective bonding conductor, is the conductor continuous and not cut at the clamp?
- Is supplementary bonding visible in the bathroom? It may not be required of course but you still need to look, as you will not have carried out a test to see if it is required yet.
- Does any equipment fitted in rooms containing a bath or a shower meet the required IP ratings?
- Is equipment fitted in bathrooms suitable for the environment? See Section 701 of BS 7671.
- Are there any socket outlets in the bathroom? If so, are they SELV or PELV? Are any 13 amp socket outlets a minimum of 3m horizontally from zone 1?
- Has any bedroom had a shower installed? If so, are the socket outlets 3m from the shower and RCD protected?
- Is there any evidence of mutual detrimental influence? Are there any cables fixed to water or gas pipes or any other non-electrical services? (The cables need to be far enough away from non-electrical services to avoid damage if the services are ever worked on.)
- Are cables of different voltage bands segregated? Low voltage, extra low voltage, telephone cables, television aerials and data cables need to be kept apart. They must not be clipped together (although they are permitted to cross).
- Where photovoltaic systems have been installed, has the inverter adequate ventilation? Usually 150mm minimum is required all round.
- Has the inverter got d.c. rated isolation on the d.c. side?
- Is the d.c. isolator correctly identified?
- Has the inverter got a lockable a.c. isolator on the a.c. side?
- Are the PV cables correctly identified? Brown for positive and grey for negative.
- Where the inverter is sited away from the consumer unit, is a lockable a.c. isolator fitted next to the consumer unit?
- Are dual isolation labels in place?

Whilst these items are being checked, look in any cupboards for sockets, lighting points or any other items of connected electrical equipment. It may be that your customer is uncomfortable with you poking around and of course this is understandable; however every effort should be made to explain to your customer how important it is to try and find everything which is connected to the fixed wiring. When it comes to the testing of the installation it may be that items not found may result in false readings, or of course if it is electronic equipment it may well be damaged by the test voltages.

It is vitally important that you document any areas that cannot be investigated; they must be recorded in the extent and limitation section of the electrical installation condition report. During this purely visual part of the inspection you will gain some information which will help you assess the condition of the installation, and indeed any alterations which have been carried out.

Providing that you are happy that the installation is safe to work on, a more detailed visual inspection can be carried out and the dreaded but necessary form filling can begin.

Once again begin at the consumer unit; wherever possible it should be isolated before you start even if this does cause some inconvenience. However, due to the type of installation, often complete isolation is not possible, particularly on some commercial or industrial installations.

In an ideal world isolation would take place before any covers are removed. If the equipment which you are working on has not been isolated, once you have removed the cover you will be working live and great care must be taken, and it is always a good idea to wear protective glasses just in case of any arcing.

When carrying out a visual inspection, your first impression will be important:

- Has care been taken over the terminations of cables (neat and not too much exposed conductor)?
- Are all of the cables terminated and all of the connections tight (no loose ends)?
- Are there any signs of overheating?
- Is there a mixture of protective devices? This will help to indicate whether there have been alterations.
- Are there any rubber cables?
- Are there any damaged cables (perished or cut)?
- Have all circuits got circuit protective conductors (CPCs)?
- Are all earthing conductors sleeved?
- Look to see if the protective devices seem suitable for the cable sizes that they are protecting.

- Make a note of any protective devices which do not seem to be right, these could be D or 4 type circuit breakers – these will require further investigation.
- Are all barriers in place? Remember the internal barriers should comply with IPXXB.
- Have all of the circuit conductors been connected in the correct sequence, with the line, neutral and CPC of circuit 1 all being in the corresponding terminal? Preferably with the highest rating circuit nearest the main switch.
- Have any protective devices got multiple conductors in them? Are they the correct size (all the same)?
- Is there only one set of tails or has another board been connected to the original board by joining at the terminals?

Having had a detailed inspection of the consumer unit, carry out a more detailed inspection on the rest of the installation. Where there is more than one consumer unit or distribution board, it will usually make the inspection easier if you deal with the installation one board at a time.

As previously mentioned the extent and limitation of the inspection must be agreed with the person ordering the work before starting, and of course each job will need to be treated differently depending on the factors surrounding the installation. For instance, how long ago was the installation last inspected, what is the installation used for, has there been a change of use since the last inspection – these along with many other factors can have an influence on the agreed percentage of inspection.

As a minimum the inspection should be 10 per cent and again this depends on the type of installation. For a domestic installation with one or perhaps two consumer units then 10 per cent of each circuit should be inspected. On an industrial or commercial installation the inspection can be limited to 10 per cent of the circuits on each board. In these cases where something is found which could have an influence on other parts of the installation, further investigation would need to be considered. It is quite possible that a further discussion with the client would be required to advise them that the percentage of the inspection would need to be increased.

During your preliminary walk around, you will have identified any areas of immediate concern and these must be addressed as your inspection progresses. However there is no reason why you cannot start some of your testing at this point.

Providing you have the past test results it is a good idea to carry out earth fault loop impedance (Z_s) tests as you work your way around the installation. Where the results are the same as the previously recorded ones then there would be no reason to carry out a dead test for continuity of circuit protective conductors.

Let's look at this in a little more detail:

With an initial verification on a circuit the first test to be carried out is continuity of the CPC. This test gives the $R_1 + R_2$ result which is recorded in the correct column of the schedule of test results. The measured value of Z_e will also have been recorded and the value of Z_s, which should be no higher than $Z_e + R_1 + R_2$, must also be entered onto the schedule.

Where we record Z_s using this method we are really recording the calculated value; if we were to measure the Z_s of the same circuit using an earth loop impedance tester, it is quite likely that the value would be lower than the calculated value. There is nothing wrong with this and it does not mean that the tester we are using is not accurate; all that is happening is that the live test is measuring through any parallel paths which exist, these parallel paths could be through any protective bonding which has been installed and possibly other parts of the earthing and bonding system.

As a rule of thumb, providing the measured value of Z_s recorded during a periodic inspection and test does not exceed the values of $Z_e + R_1 + R_2$ from the past test results, the circuit will not need to have an $R_1 + R_2$ test carried out on it.

It may be of course that the new measured value of Z_s is higher than the original measured value of Z_s. This could be because a parallel path has been removed. A metal water main being replaced with a plastic one would result in the measured value of Z_s increasing, but it will not increase above the original measured values of $Z_e + R_1 + R_2$ – if it does then further investigation would be required. If the high reading was consistent over all of the circuits then the first check would be the accuracy of the instrument; after that if the instrument is accurate check the main earthing terminal and the earthing conductor.

What are we looking for during a periodic inspection?

Remember the installation may not comply with the current regulations, but it is being inspected and tested against the current regulations. For that reason many installations will not meet the current standards, although clearly an installation which was safe when it was installed does not become unsafe because the regulations have been updated. In these cases a judgement has to be made by you as to whether the installation is recorded as satisfactory or unsatisfactory.

As an example, compliance with BS 7671: 2018 will require that in most instances circuits have RCD protection. It will be many years before this will be found to be the case when carrying out a periodic inspection. In most instances this type of non-compliance would be recorded as C3 and the installation could be deemed to be satisfactory when completing the electrical installation condition report.

Let's look at a selection of circuits.

Shower circuit

- Has it been provided with additional protection (30mA RCD)?
- Is isolation provided? If so, is it within the prescribed zones? (Remember the switch can be anywhere outside of zone 2.)
- Has the correct cable/protective device been selected? There are two important things to remember here. First, is the cable size as recommended by the shower manufacturer? For the larger type of shower it is usual for the recommended size to be 10mm^2. Second, for this type of circuit it is possible to have a cable which has a current rating of less than the rating of the protective device. This is because we do not have to provide overload protection to circuits which cannot overload.
- Has supplementary bonding been provided if required?
- Is the shower unit secure?
- Is there any evidence of damage?
- Are the connections tight?
- Is there any evidence of water ingress?
- Do the conductors show any evidence of overload?
- Is the shower in a bedroom? If so, are all 13A socket outlets a minimum of 3m from the edge of the shower and RCD protected?

Cooker circuit

- Is there any evidence of damage?
- Is the switch within 2m of the cooker or hob?
- Has the cooker switch got a 13 amp socket outlet? If so, it requires a 0.4 second disconnection time and RCD protection (note 1).
- Has green and yellow earth sleeving been fitted?
- If it is a metal faceplate, has it got an earth tail fitted between the plate and the metal mounting box?
- Is the cable the correct size for the protective device?
- Are there any signs of overloading?
- Is the cooker outlet too close to the sink? Building regulations require that any electrical outlet installed after January 2005 should be at least 300mm from the sink.

Note 1. For any installation carried out after June 2018 any final circuit with a protective device rating of up to and including 63A for socket outlets and 32A for fixed equipment must have a minimum disconnection time of 0.4 seconds if connected to a TN system and 0.2 seconds if connected to a TT system.

Socket outlets

- Are there any signs of damage or overload?
- Are all socket outlets secure?
- Is there correct coordination between protective devices and conductors?
- Do any metal socket outlets have an earth tail between the box and the metal faceplate?
- Are the cables throughout the circuit the same size?
- Are outside socket outlets watertight? It is always a good idea to have a good visual inspection of these.
- Are there any outlets in a room containing a bath or shower? If there is, are they SELV or are they at least 3 metres from the edge of the bath or shower and are they 30mA RCD protected?
- Are all of the socket outlets RCD protected? This is a requirement for any sockets installed after 2008. Where the installation is pre-2008 the non-compliance should be recorded as a C3.
- Are there any outlets within 300mm of a sink?

Fused connection units and other outlets

- Are they in a bathroom? If so, are they in the correct zones?
- Are they securely fixed?
- Are they fitted with the correct size fuse?
- Functional switching devices shall be suitable for the most onerous duty that they are intended to perform (537.3.1.2).

Immersion heater circuits

- Is there correct coordination between the protective device and live conductors? Again you have to be careful here as an immersion heater cannot overload and for that reason it does not need overload protection. It would be perfectly acceptable for a cable supplying a 3kW immersion heater, or any fixed resistive load, to be connected to a protective device with a higher rating than the cable. This is of course providing it met with the voltage drop requirements and Z_s values required for automatic disconnection.
- Has the CPC been sleeved?
- Is the immersion heater the only equipment connected to this circuit? (Any water heater with a capacity of 15 litres or more must

have its own circuit – *On-site Guide* Appendix H). It is not unusual to find that the immersion heater point is also used to supply the central heating controls, particularly on an older installation. I have usually managed to gain compliance by disconnecting the immersion heater. In most installations with central heating the immersion heater is only used as an emergency backup. Don't disconnect it without asking the customer though.

- Has the immersion heater been connected with heat resistant cord?
- The immersion heater switch should ideally be a 20 double pole switch, although some electricians use a fused connection unit fitted with a 13a fuse. This is fine but in reality the fuse is operating at its maximum, and I have lost count of the number which I have seen burnt out after a period of time.
- The immersion heater should never be connected by a plug and socket.
- If the protective supplementary bonding for the bathroom has been carried out in the airing cupboard, does the bonding include the immersion heater switch? (It should.)
- Is the immersion the type with an overheat protection cut out provided? It should be.

Lighting circuits

- Is there correct coordination between the protective device and the live conductors?
- How many points are there on the circuit? A minimum rating of 100 watts must be allowed for each outlet. Shaver points, clock points and bell transformers may be neglected for load calculation. As a general rule a domestic installation should have no more than 10 lighting points per circuit; this is because single lamp fitting can be changed for a three lamp fitting with little effort, it does not take many of these types of exchanges to overload a circuit.
- A commercial installation generally consists of known loads such as fluorescent fittings or discharge lamps; for this type of circuit the load can be calculated just as any other circuit would be. Of course we must remember that these types of lamps are rated by their output, and unless they are power factor corrected the lamp output must be multiplied by a factor of 1.8.
- Are all switch returns colour identified at each end?
- Have the switch drops got CPCs? If they have, are they sleeved green and yellow?
- Are the switch boxes made of wood or metal?
- Are the ceiling roses suitable for the mass hanging from them?
- Only one flexible cord is permitted to be connected to each ceiling rose, unless they are designed for multiple cords.

- Are domestic lighting circuits RCD protected (411.3.4)?
- Light fittings in bathrooms must be suitable for the zones in which they are fitted.
- Are luminaires fitted a suitable distance from combustible surfaces?
- Are luminaires appropriate for the location?
- Are luminaires suitable for the surface to which they are fixed?
- Is the line conductor to an ES lampholder connected to the centre pin? This does not apply to E14 and E27 lampholders as they are all insulated (Regulation 643.6).

Three phase circuits/systems

These circuits must be inspected for the same defects that you could find in other circuits. In addition to this:

- Are warning labels fitted where the voltage will be higher than expected? For example, a lighting circuit with more than one phase in it, or perhaps where socket outlets which are close to each other are one different phases.
- Are the conductors in the same sequence right through the installation?
- Remember when measuring I_{pf} the value recorded should be double the measured line to neutral current.

Occasionally other types of installations or circuits will be found. In these cases the same common sense approach to inspection should be applied. Always remember that all you are doing is checking to ensure that the installation is safe for continued use.

CHAPTER 4

Periodic testing

The level of testing required for an electrical installation condition report will usually be far less than that required for initial verification; this is providing of course that previous inspection and test documentation is available. If it is not, then it will be necessary to carry out a full survey, and the complete range of tests must be carried out on the installation. This will be necessary to provide circuit charts and a comprehensive set of test results.

The level of testing will depend largely on what the inspector discovers during the visual inspection, and the value of any test results obtained while carrying out sample testing. If any tests show significantly differing results from previously recorded results for no apparent reason, then further tests may need to be carried out.

In some cases, up to 100 per cent of the installation will need to be tested, particularly where the past documentation is not available. Periodic inspecting and testing can be dangerous, and due consideration must be given to safety.

Persons carrying out the inspection and testing must be competent and experienced in the type of installation being inspected and also in the use of the test instruments being used.

When carrying out a periodic inspection and test it can be expected that the installation is live and is being used; for this reason it is not required that the tests are carried out in any set sequence for the completion of the condition report and the sequence of tests is left to the person carrying out the inspection and test to decide upon.

As it is usual for the installation to be live, the first test which I normally carry out is an earth loop impedance test close to the origin of the supply. This is just to ensure that there is in fact an earth on the installation and that the polarity is correct before I start.

Practical Guide to Inspection, Testing and Certification. 978-1-138-61332-4

As previously mentioned it is down to the person carrying out the inspection to decide on the level of testing required when test results are available; where they are not available the whole installation must be tested wherever possible.

Where past documentation is available it is perfectly acceptable to transfer some values from the original documents over to the new documents without carrying out the test; this is a common sense approach to testing.

It is made quite clear in BS 7671: 2018 Regulation 652.2 that a periodic inspection should comprise of a detailed examination supplemented by appropriate tests to prove the requirements for disconnection times are complied with.

On the original documentation there should be the value of Z_e for the supply and the value of $R_1 + R_2$ for each circuit. Also recorded should be the measured value of Z_s, although often the value of Z_s is a calculated value using the formula $Z_s = Z_e + R_1 + R_2$.

During periodic inspections and tests the value of Z_s for each circuit must be measured; the highest value will be found at the furthest point of the circuit and this will be the value which will need to be measured. If we know the installation well then the furthest point will be known to us, if not it may be necessary to test at several points on the circuit to find it.

Once found, the measured value of Z_s can be compared to the value recorded on the original or last schedule of test results. If it is the same then all well and good; however sometimes it will not be the same and this could be for various reasons.

The original recorded value could have been the calculated one or a parallel path could have been removed or added. The removal of a parallel path (which could be as simple as a metal water main being changed for a plastic one) could increase the measured value. The introduction of a parallel path (possibly something like an oil line being installed for a boiler or some structural steel in contact with the earth) would lower the measured value.

In these situations all that is required is for the original recorded values of Z_e and $R_1 + R_2$ to be added together. Providing the result is not higher that the value which has just been measured we can simply transfer the original values of Z_e and $R_1 + R_2$ onto our new certificate as it is unlikely that they have changed.

If the value is higher, then it is always worth checking the integrity of the connections of the earthing conductor, or the accuracy of the test instrument particularly if high values are being measured on all of the circuits.

Providing the earth loop resistance test proves satisfactory there is generally no real requirement for further tests to be carried out. This is of course down to the inspector and his/her general view on the condition of the installation. None of this is cast in stone and each installation must be treated on its own merits.

Table 4.1 shows the recommended tests when further testing is required.

Table 4.1 Recommended tests or further testing

Continuity of protective bonding conductors	All main bonding and supplementary bonding conductors
Continuity of circuit protective bonding conductors	Between the distribution board earth terminal and exposed conductive parts of current-using equipment Earth terminals of socket outlets and fused connection units (test to the fixing screw of the accessory for convenience)
Ring circuit continuity	Only required where alterations or additions have been made to the ring circuit or where the measured Z_s value is not compatible with original measured values
Insulation resistance	Only between live conductors joined and earth, unless you can be sure that nothing is connected to the circuit. If testing a lighting circuit the test can be carried out between live conductors providing the functional switch is open.
Polarity	Live polarity at the origin of the installation Socket outlets At the end of radial circuits Distribution boards
Earth electrode	Isolate the installation and disconnect the earthing conductor from main earth terminal to avoid parallel paths
Earth fault loop impedance	At the origin of the installation for Z_e Distribution boards for the Z_e of that board Socket outlets and the end of all other circuits to check Z_s values
Functional tests	RCD test and manual operation of switches and isolators
Voltage drop	Calculate using tables or $R_1 + R_2$ values

Video footage is also available on the companion website for this book.

Voltage drop and inspection and testing

Regulation 525.1 states:

In the absence of any other consideration, under normal service conditions the voltage at the terminals of any fixed current-using equipment shall be greater than the lower limit corresponding to the product standard relevant to the equipment.

In the UK we are more specific or, to put it another way, we mention safety and voltage drop tolerances. See Regulations 525.201 and 202; these two regulations direct you to Page 338 in BS 7671, Appendix 4.

In other words we need equipment connected to circuits to work both correctly and safely.

Verification of voltage drop

Then we turn to Regulation 643.11, which tells us how voltage drop may be evaluated (worked out). It also has a note at the foot of the regulation that states 'Verification of voltage drop is not normally required during initial verification'.

So if we assume that voltage drop has been resolved/confirmed during the design and circuit lengths and loads confirmed during initial verification, then this is true.

This only leaves verification of voltage drop for periodic inspection and testing. So, how do we check this?

The easiest way to look at this is Ohm's law.

$$\text{Voltage} = \text{Current} \times \text{Resistance}$$

Voltage is the voltage drop you are looking for.
Current is either the design current of the circuit ($I = P/V$) or the size of the protective device.
Resistance is the value of the line r_1 and neutral r_n.

We need to consider one more thing, and that is the effect of temperature on the conductors. The resistance of conductors from the *On-site Guide* are given at 20°C and when we are testing, the conductors are highly unlikely to be at 70°C. Table I3 of the *On-site Guide* provides us with the information required. Are the cables bunched? Are they thermoplastic or thermosetting? Are they 70°C or 90°C?

So if we take a 70°C thermoplastic cable, as an example, incorporated or bunched, then we need to apply a factor of 1.2. So the way we calculate voltage drop on an existing circuit is:

Voltage drop = Current × Resistance × Multiplier for operating temperatures

If we have a circuit with a combined line and neutral resistance of 0.31Ω and a protective load current of say 45A we would determine the value of voltage drop like this.

$$V = I \times R \times \text{Temp factor}$$
$$V = 45 \times 0.31 \times 1.2 = 16.74 \text{ Volts}$$

In Appendix 4 Section 6.4 of BS 7671 Table 4Ab shows us that voltage drop tolerances of 3 per cent for lighting circuits and 5 per cent for other circuits are the maximum allowable voltage drop between the origin of an installation and any load point, when the supply is a public low voltage distribution system.

If we assume the supply is 230V we can say that 3 per cent of 230V is 6.9 Volts and that 5 per cent of 230V is 11.5 Volts.

We can clearly see from the example above that the circuit would fail on voltage drop as it higher than either of the maximum available tolerances' for of a *'lighting circuit'* or *'other uses'*.

Voltage drop in conductors

Part of the periodic inspection process requires that we check each circuit for current carrying capacity and voltage drop. To check the suitability of the current carrying capacity, it is simply a matter of looking at the installation method and then checking the current rating for the cable using the tables in Appendix 4 of BS 7671.

To ensure that a circuit meets the requirements for voltage drop is slightly more complex. It is a very difficult task to accurately measure volt drop in a circuit due to loads and fluctuating supply voltages. It is also not a requirement of BS 7671 that it is carried out by voltage measurement.

The most practical method is to use the resistance of the line and neutral of the circuit and calculate the volt drop. We can measure the resistance of R_1 and R_n and then multiply the resistance by the current in the circuit. This will give us the volt drop which must be multiplied bt 1.2 to compensate for the operating temperature of the cable.

Example 1

A circuit is wired in 2.5mm² / 1.5mm² twin and earth 70°C PVC cable which is 21 metres in length. The current in the circuit is 17 amps.

The measured value of resistance is 0.31Ω.

$$\text{Volt drop} = I \times R$$

$$17 \times 0.31 = 5.27v$$

We must now multiply the answer by 1.2 – this will compensate for the temperature rise when the cable is under load.

$$5.27 \times 1.2 = 6.32v$$

This is the voltage drop for the circuit.

Of course the calculation would require the $R_1 + R_n$ values to be recorded and very often this would be impractical, particularly whilst carrying out a periodic inspection as it would probably require the disconnection of parts of the installation. A much better method is to use the values of $R_1 + R_2$ which will already have been measured and recorded.

Example 2

Using the circuit from example 1 the $R_1 + R_2$ value will be 0.4Ω.

$$R_1 + R_2 \times I = V$$

$$0.4 \times 17 \times 1.2 = 8.16 \text{ v}$$

This calculation is using the resistance of the CPC and the line conductor. Of course, the CPC will have a higher resistance and will not give an accurate value of volt drop. The regulations only require that we check that the value of volt drop is below the permitted maximum.

Providing this is not a lighting circuit the calculated voltage drop would be satisfactory. If a more accurate value is required, or perhaps the method used above resulted in a value which looked too high, then volt drop can be calculated using this method:

$$\frac{CSA_{line}}{CSA_{line} + CSA_{cpe}} \times (R_1 + R_2) = R_2$$

$$\frac{2.5}{2.5 + 1.5} = 0.625$$

$$0.625 \times 0.4 = 0.25\Omega$$

0.25 is the resistance of R_2; to find the resistance of R_1 we must subtract R_2 from $R_1 + R_2$:

$$R_1 = 0.4 - 0.25 = 0.15$$

Now we have the resistance of R_1 we need to double it as the current flows in both R_1 and R_n:

$$0.15 + 0.15 = 0.3\Omega$$

Now we have the value of $R_1 + R_n$ we can use our original calculation to find an accurate value of volt drop.

$$R \times I \times 1.2 = \text{volt drop at } 70°C$$

$$0.3 \times 17 \times 1.2 = 6.12v$$

As you can see this is only 0.2 volts different from the original calculation which is plenty accurate enough for this type of calculation.

Testing of electrical installations

Safe isolation

The importance of carrying out safe isolation correctly cannot be overemphasised; it must be carried out in the correct sequence which must be repeated each time a single circuit or a complete installation is to be isolated. If the same procedure is carried out each time it will soon become a habit, a habit which prevents you, and others, from being killed or injured from electric shock.

To carry out electrical isolation safely it is vital that not only are the correct procedures followed but also that the correct equipment is used. The Health and Safety Executive have produced a document known as GS38, which provides guidance on the type and use of test equipment used for measuring voltage and current, particularly leads and probes.

GS38 is not a statutory document; however, if the guidance given is followed it will normally be enough to comply with the Health and Safety at Work Act 1974 and the Electricity at Work Regulations 1989, along with any other statutory requirements which may apply.

To carry out safe isolation correctly we must have available the correct equipment and tools which will include:

- an approved voltage indicator or test lamp (Figure 4.1)
- warning notices (Figure 4.2)
- locking devices (Figure 4.3)
- a proving unit (Figure 4.4).

Another very useful piece of equipment is an R_1 and R_2 box (Figure 4.5). This will not only be useful for the safe isolation of socket outlets, it can also be used for ring circuit testing, or the testing of

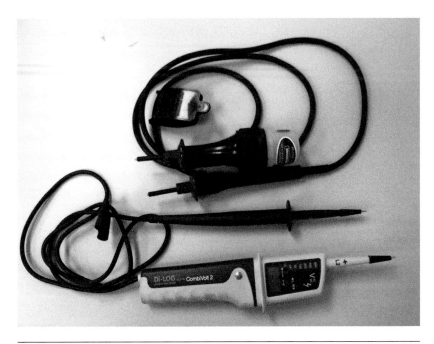

Figure 4.1 Approved voltage indicator and test lamp

Figure 4.2 Warning notices

Figure 4.3 Locking devices

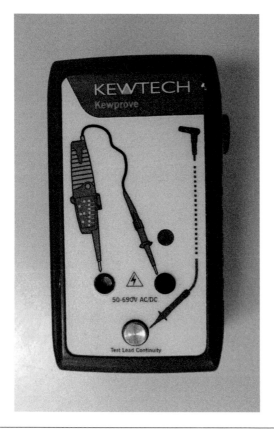

Figure 4.4 Proving unit

Figure 4.5 $R_1 + R_2$ box

any radial circuits incorporating a socket or socket outlets without having to remove them from the wall.

The leads of test equipment used to test voltages above 50v should be:

- flexible and long enough, but not so long they become difficult to use
- insulated to suit the voltage at which they are to be used
- coloured where necessary to identify one lead from another
- undamaged and sheathed to protect them from mechanical damage.

The probes should:

- have a maximum of 4mm exposed tip (preferably 2mm)
- be fused at 500mA or be fitted with current limiting resistors
- have finger guards (to prevent fingers from slipping onto live terminals)
- be colour identified.

Under no circumstances should a multimeter, a volt stick or a neon indicating screwdriver ever be used for safe isolation.

Safe isolation procedure

It is very important to ensure that the circuit which you are intending to isolate is live before you start. This will also be an opportunity to check the correct operation of the voltage indicator or test lamp.

Step 1

Ensure the circuit is live and that the voltage indicator is working (Figure 4.6).

Be careful! Most test lamps will trip an RCD when testing between live and earth. It is better to use an approved voltage indicator to GS38 as most of these will not trip RCDs.

If the circuit appears to be dead, you need to know why before proceeding.

- Is somebody else working on it?
- Is the circuit faulty?
- Is it connected?
- Has there been a power cut?
- Is the voltage indicator or test lamp working?

You must make absolutely certain that you and you alone are in control of the circuit to be worked on. Once you are sure that the

Safe isolation procedure

Video footage is also available on the companion website for this book.

Figure 4.6 Test line to neutral

circuit is live and that your voltage indicator is working you can proceed with the isolation as follows.

Step 2
Test between all live conductors and earth (Figure 4.7).

Step 3
Locate the point of isolation, isolate and lock off. Once isolated place a warning notice (*Danger Electrician At Work*) at the point of isolation (Figure 4.8).

Step 4
Test the circuit to prove that you have isolated the correct circuit (Figure 4.9).

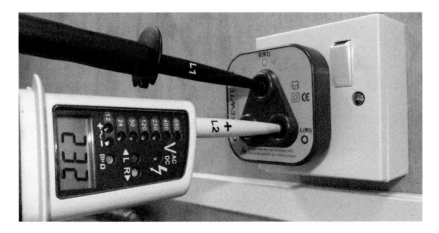

Figure 4.7a Test line to earth

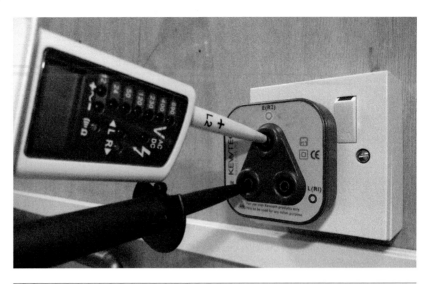

Figure 4.7b Neutral to earth

Figure 4.8 Locked off

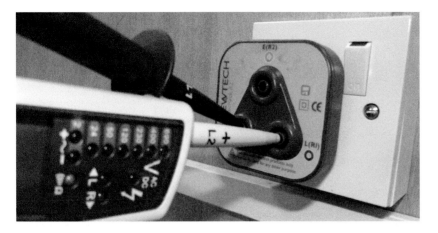

Figure 4.9a Test line to neutral

Figure 4.9b Line to earth

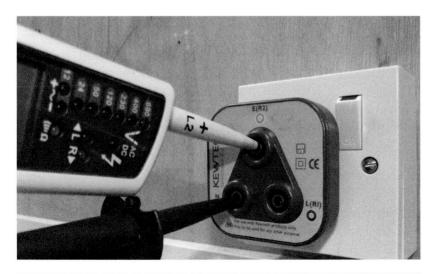

Figure 4.9c Neutral to earth

Figure 4.10 Retest device

Figure 4.11 Double check

Step 5

Check that the voltage indicator is working by testing on a known supply or proving unit (Figure 4.10).

Step 6

To be on the safe side I always just check the circuit is dead for a second time (Figure 4.11). Better to be safe than sorry!

When carrying out safe isolation never assume anything, always follow the same procedure.

If the circuit which has been isolated is to be disconnected, always try to isolate the consumer unit completely. Sometimes this can prove difficult for various reasons but it is much safer.

 Main protective bonding

 Video footage is also available on the companion website for this book.

Testing of protective bonding conductors

Main protective bonding

This test is carried out to ensure that the protective bonding conductors are unbroken and have resistance low enough to satisfy the requirements of BS 7671. The purpose of the protective bonding is to ensure that under fault conditions a dangerous potential will not occur between earthed metalwork (*exposed conductive parts*) and other metalwork (*extraneous conductive parts*) in a building. Where the protective bonding is visible in its entirety a visual inspection will be suitable.

It is not the purpose of this test to ensure a good earth path, it is to ensure that in the event of a fault the exposed and extraneous conductive parts will rise to the same potential, hence the term 'equipotential bonding'. In order to achieve this it is recommended that the resistance of the bonding conductors does not exceed 0.05Ω. It should always be remembered however that many low resistance ohm meters are not accurate below 0.2Ω, and that providing the conductor connections are tight a reading of *around* 0.05Ω will be acceptable.

Chapter 54 of BS 7671 covers the requirements of protective bonding. Chapter 4 of the *On-site Guide* is also useful.

In general the maximum length of copper protective bonding conductor before 0.05Ω is exceeded is shown in Table 4.2.

Where the whole length of the conductor is not visible a test must be carried out with a low resistance ohm meter. This test can often only be carried out during initial verification; this is because one end of the bonding conductor must be disconnected to avoid the measurement including parallel paths. When disconnecting bonding it is important that the installation is isolated from the supply. On larger installations it is often impossible to isolate the complete installation and therefore the conductor must remain connected. The low resistance ohm meter must be set on the lowest possible value of ohms and the leads must be nulled or the instrument zeroed.

Step 1

Isolate the supply following the safe isolation procedure (Figure 4.12).

Step 2

Disconnect one end of the protective bonding (Figure 4.13). (If possible disconnect at the consumer unit and test from the disconnected end and the metalwork which the bonding is connected to. This will test the integrity of the bonding clamp.)

Table 4.2 Maximum length of copper protective bonding conductors

Size mm^2	Length in metres
10	27
16	43
25	68
35	95

Figure 4.12 Isolate the supply

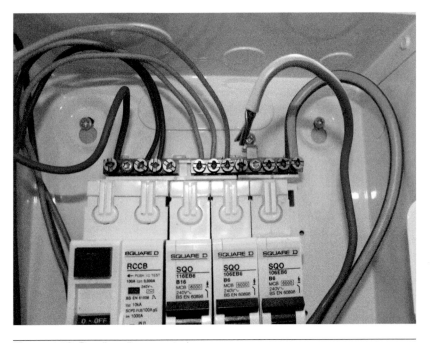

Figure 4.13 Bonding disconnected

Step 3

Null the leads or record their resistance value (Figure 4.14). (The leads may be long as the only way to measure a bonding conductor is using method 2 which is from end to end.)

Step 4

Connect one lead to the disconnected conductor (Figure 4.15).

Step 5

Connect the other lead to the metal work close to the bonding clamp (Figure 4.16).

Step 6

If the instrument is not nulled remember to subtract the resistance of the leads from the total measured resistance. This will give you the resistance of the bonding conductor. If the meter and leads have been nulled then the value measured will be the resistance of the bonding conductor (Figure 4.17).

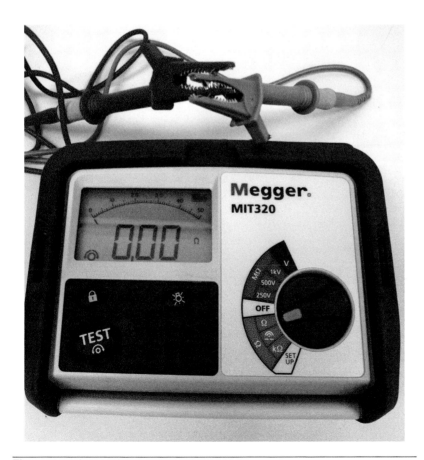

Figure 4.14 Low resistance ohm meter

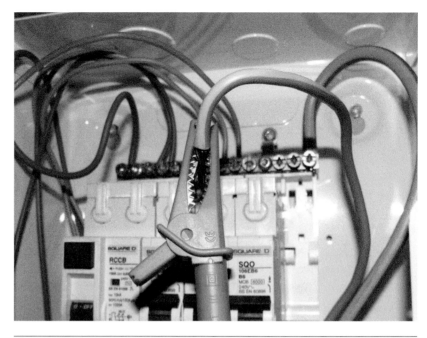

Figure 4.15 Lead connected

Figure 4.16 Second lead connected

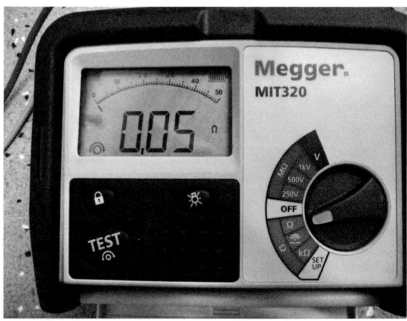

Figure 4.17 Resistance value

Step 7

Ensure that the bonding conductor is reconnected on the completion of the test (Figure 4.18).

While carrying out this test it is a good opportunity to check that the correct type of bonding clamp has been used (it must be to BS 951) and that a label is present. Where the protective bonding is looping between services it must not be cut at the bonding clamp (Figures 4.19 and 4.20).

If the installation cannot be isolated it is still a good idea to carry out a test. The resistance must be no greater than a maximum of 0.05Ω, as any parallel paths will result in the resistance measurement being lower. A measured resistance of greater than 0.05Ω must be reported as unsatisfactory and requiring improvement.

Remember that where the protective bonding is visible over its entire length, a visual inspection would be satisfactory although consideration must be given to its length.

For recording purposes on inspection and test certificates and reports, no value is required but verification of its size and suitability is.

Items requiring bonding would include any incoming services, such as water, gas, oil and LPG as well as structural steelwork, the central heating system, air conditioning and lightning conductors within an installation (*bonding to lightning conductors must comply with BS EN 62305*). It is always advisable to seek the advice of a lightning conductor specialist before connecting any kind of protective bonding to a lightning protection system.

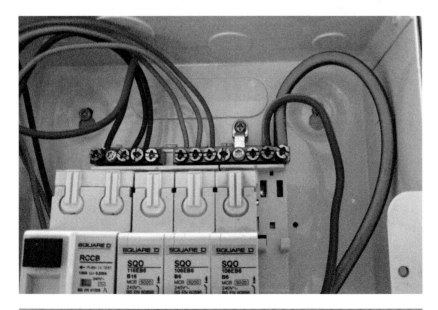

Figure 4.18 Reconnect bonding

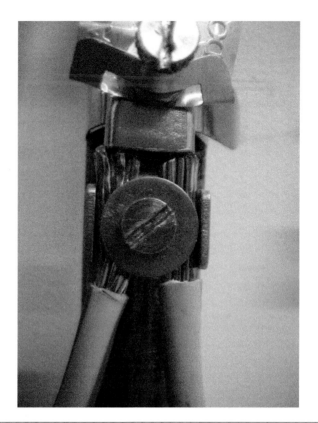

Figure 4.19 Incorrect!

Figure 4.20 Correct!

The list of parts to be bonded is not a concise list and consideration should be given to bonding any metalwork that could introduce a potential within a building.

Continuity of protective supplementary bonding conductors

There are two general reasons for providing protective supplementary bonding; these are where disconnection times cannot be met (Regulation 419.3) or in areas where there is an increased risk of electric shock. This would be classed as additional protection (Regulation 415.2.1), generally these areas would be rooms containing a bath or a shower, swimming pools and other special locations.

Where protective supplementary bonding is required and any doubt about its effectiveness exists, Regulation 415.2.2 requires that it is checked; in these situations a simple test is required.

Where automatic disconnection is met by using a circuit breaker or fuse, the regulations provide a formula $R < 50V/I_a$. Where the circuit is protected by an RCD, the formula becomes $R < 50V/I_{\Delta n}$ (in some special locations 50v must be substituted by 25v).

Where this calculation is used, it will ensure that any touch voltage will not rise above 50v before the protective device operates.

To use the formula where a fuse or circuit breaker is used for automatic disconnection of supply (ADS), the first step is to find the current which will operate the protective device (I_a) within a maximum of 5 seconds.

Let's say that the protective device is a 20A BS 3036 rewirable fuse. To find the current which will automatically operate this device in 5 seconds we can look in Appendix 3 of BS 7671. Fig 3A2(b) is the table which we need to look at. This shows that the current required to operate the fuse in 5 seconds is 60A.

The value can also be found by using the maximum Z_s value for a 20A fuse which can be found in Table 41.4 in Appendix 3. The value is 3.64Ω.

The calculation can now be carried out:

$$I_a = \frac{230}{Z_s} \times 0.95 \text{ (referred to as } C_{min})$$

$$\frac{230}{3.64} \times 0.95 = 60A$$

$$I_a = 60A$$

(*This method can be used for all protective devices.*)

This value can now be used to verify if the area requires protective supplementary bonding or not. The calculation is:

$$R = \frac{50v}{I_a}$$

$$R = \frac{50}{60}$$

$$R = 0.83\Omega$$

The maximum permitted value between exposed or extraneous conductive parts in the area is 0.83Ω. If the measured resistance is higher than 0.83Ω then protective supplementary bonding will be required.

The resistance values used will be different depending on the type, and rating of the protective device which is being used for protection of the circuit.

In areas where an RCD is being used for protection, the following calculation must be used to find the maximum resistance permitted between exposed and extraneous conductive parts before protective bonding is required.

$$R = \frac{50v}{I_{\Delta n}}$$

The trip rating of the RCD is $I_{\Delta n}$. For example if it is 30mA then the calculation is:

$$R = \frac{50}{0.03}$$

$$R = 1666\Omega$$

Providing the resistance between parts is 1666Ω or less, protective supplementary bonding would not be required.

Apart from areas where automatic disconnection times cannot be met and special locations, due consideration must be given to other areas where there is an increased risk of electric shock. There is no specific requirement to install protective supplementary bonding in kitchens; however if it is thought that there is an increased risk of electric shock (perhaps a commercial kitchen where there is a lot of stainless steel tables and worktops), there is no reason at all why it cannot be installed. It will do no harm provided it is installed correctly.

It is not unusual to see supplementary bonding in places where it is not really required. This is often for one of two reasons: first, that the installer/designer does not really understand bonding and second that it has been installed for visual purposes only.

Many people will expect to see bonding; most of them will not understand what it is for and it is called all sorts of names. I am sure that most of us have heard it called 'cross bonding' or 'earth bonding'. But for whatever reason if the customer expects to see it then it is often easier to install it than argue about it. If a quiet life can be had by installing a couple of earth clamps and a short length of 4mm^2 then in most cases it will be worthwhile.

Wherever protective supplementary bonding is installed, a test must be carried out to ensure that the resistance between the bonded parts is equal to or less than the value obtained by using the calculation described earlier. The instrument used for this is a low resistance ohm meter; a visual check must also be made to ensure that the earth clamps are to BS 951 and that the correct labels are in place.

It is perfectly acceptable to utilise any metal pipe work and structural steel within an area as bonding conductors. The bonding can also be carried adjacent to the area providing the integrity of the pipe/steel work can be assured. An airing cupboard would be a good example of a suitable place to bond where bonding is required in a bathroom.

Bonding can also be carried out using pipe work within the roof space. This can be useful where perhaps an electric shower or a lighting point requires bonding. Providing the pipe work is bonded elsewhere in the building it is perfectly acceptable just to bond from the shower or the lighting point on to the nearest pipe. If in doubt a test should be carried out to ensure that the pipe is bonded and of course the correct clamps and labels must be used.

When pipe work is used as protective supplementary bonding, it must always be tested to ensure that the resistance between any exposed or extraneous conductive part does not exceed that which has been calculated using the formula described earlier. This test is very easy to perform using a low resistance ohm meter, possibly with long leads. Before performing the test it is important to check the meter for any damage, correct operation and accuracy and of course the leads must be nulled. The probe of one lead must be placed onto one metal part and the other lead on another metal part (Figures 4.21 and 4.22).

Video footage is also available on the companion website for this book

When using metal pipe work as bonding conductor's problems can arise where the pipe work is altered and plastic push fittings are used. Clearly these fittings will not conduct, and contrary to popular belief neither will the water inside the fitting and the bonding continuity will be affected. Wherever plastic fittings are used consideration must always be given to providing bonding across the fitting (Figure 4.23).

It is a common belief that water in pipe work will conduct: in fact the current which will flow through water across a plastic fitting filled with water is very small.

Figure 4.21 Lead touching tap

Figure 4.22 Lead on unpainted part of radiator

Figure 4.23 Fitting bonded across

Figure 4.24 Current flow through pipe

To find out how much current would flow I carried out a controlled experiment using two short lengths of copper pipe. These were joined using a plastic push fit coupler. Once fully pushed home the pipes were no more than 2mm apart (Figure 4.24).

The pipe was then filled with tap water and the ends were connected to a 230v supply and the current flowing was measured at 0.003 amperes (3mA) using a clamp meter. The current flow would increase if the water had central heating additives in it, but not considerably.

Supplementary bonding is installed where there is a risk of simultaneous contact with any extraneous and exposed conductive parts. Its purpose is to ensure that the potential difference does not rise above a safe value. In most cases this value is 50 volts, although in some special locations this value can be as low as 25 volts. These locations are described in Part 6 of BS 7671.

Determining if a metal part is extraneous or just a piece of metal

Very often it is impossible to tell whether a metal part is extraneous or not. In these situations a test should be made using an insulation resistance tester set on MΩ with a voltage of 500 volts d.c.

One lead from the tester must be connected to a known earth and the other lead connected to the metal part. A test should then be made to measure the resistance; if the value is less than 0.02MΩ (20,000Ω) then bonding is required as the part would be deemed an extraneous conductive part. Where the resistance is found to be above 0.02MΩ it is just a piece of metal and it will not need to be bonded.

If we use Ohm's law we can see how this works:

$$\frac{V}{R} = I : \quad \frac{500}{20,000} = 0.025A$$

This shows us that a current of 25mA could flow between conductive parts; this of course is at 500 V, if the fault was on a 230v supply then the current flow would be half, which would be 12.5 mA (0.012A). A current of this magnitude is unlikely to give a fatal electric shock although it will be very painful.

This test should not be confused with a continuity test; it is important that an insulation resistance tester is used.

Continuity of circuit protective conductors

This test is carried out to ensure that the circuit protective conductors of radial circuits are intact and continuous throughout the circuit. The instrument used for this test is a low resistance ohm meter which should be set on the lowest value possible.

This is a dead test and it must be carried out on an isolated circuit.

There are two methods which can be used to perform this test.

Method 1

Step 1

Using a short lead with a crocodile clip on each end connect line and CPC together at one end of the circuit (Figure 4.25). (It does not matter which end the connection is made, although it is often easier to connect at the consumer unit as it will certainly be one end of the circuit.)

Don't forget! When nulling the resistance of the test leads, the lead used as a link must also be nulled before the test is carried out.

R$_1$ and R$_2$ test

 Video footage is also available on the companion website for this book.

Step 2

At each point of the circuit, test between line and CPC (Figure 4.26).

Keep a mental note of the measured value as you carry out the test, the highest value measured will be recorded as the R$_1$ + R$_2$ value for the circuit. This value must be recorded on the schedule of test results.

Where the highest measured value is clearly not the furthest point of the circuit, further investigation must be carried out as the measured value may be due to a loose connection (high resistance joint).

In some instances the value of R$_2$ may be required, obviously if the live conductors are the same size as the CPC, the R$_1$ + R$_2$ value will simply just need to be halved. This is because the conductors are all

Three phase R$_1$ and R$_2$ test

Video footage is also available on the companion website for this book.

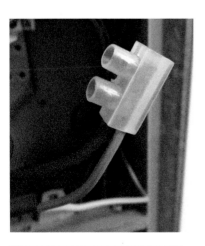

Figure 4.25 Cables joined

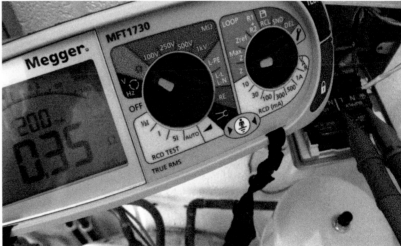

Figure 4.26 Probes on line and earth

the same CSA. In instances where the CPC is smaller than the live conductors, in a twin and earth cable for example, the resistance of the CPC (R_2) can be found by using the following calculation:

$$\frac{csa\ line}{csa\ cpc + csa\ Line} \times R_1 + R_2 = R_2$$

Example 3

A radial circuit is wired in twin and earth cable which has 2.5mm² line conductors and a 1.5mm² CPC, a test resistance of $R_1 + R_2$ was measured at 0.53Ω.

To calculate the resistance of the CPC on its own:

$$\frac{2.5}{2.5 + 1.5} \times 0.53 = 0.33\Omega \ \text{this is the value of } R_2$$

Where the CPC is smaller than the line conductor the resistance of the CPC will always be the highest.

Method 2

This method can be used where only R_2 is required and is also normally used to verify the continuity of protective bonding conductors. Method 2 is often referred to as the long lead method.

This method requires the use of a low resistance ohm meter and a long lead. The lead must be nulled before carrying out the test to ensure that an accurate value of R_2 is measured.

One lead should be connected to the earthing terminal at the consumer unit (Figure 4.27) and the other lead touched onto earthed metal at each point (Figure 4.28). This will check that each point has an earth connected and the highest reading will be the R_2 value for the circuit.

Where this method is used to check the integrity of the main protective bonding, one end of the bonding must be disconnected first.

Ring final circuit test

The purpose of the test is to ensure that:

- The conductors form a complete ring.
- There are not any interconnections.

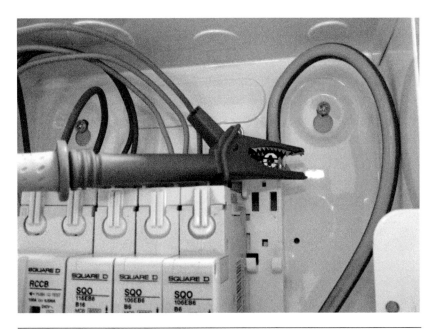

Figure 4.27 Lead connected to earthing terminal

Figure 4.28 Lead touching earthed metal

On completion of the test the polarity at each socket will also be confirmed.

When this test is carried out correctly it will also provide you with the $R_1 + R_2$ value for the circuit and it will also identify spurs.

Appendix 15, Table 15a of BS 7671 provides information on the wiring of ring final circuits. A ring circuit must be wired in 2.5mm² live conductors with a 1.5mm² CPC as a minimum size. This type of circuit can be protected by a maximum of 30/32 amp fuses or circuit breakers.

Once a ring circuit has been installed it must be tested to verify that all of the conductors form a complete loop; if they do not, then overloading of the circuit is a possibility. Where there is more than one ring circuit, there is always a possibility that the ends of the cables may get muddled and interconnected. This will result in the circuit being protected by two protective devices, which of course will mean that the circuits will be overprotected, and that isolation will only be possible by the removal or turning off of both devices.

The whole point of a ring circuit is that it can be wired in cables with a relatively small cross-sectional area, and still carry a reasonably high load current. This is because the 2.5mm² conductors are in parallel (Regulation 433.4). If we look in Table 4D5 in Appendix 4 of BS 7671 it shows the current carrying capacity for 2.5mm² cable as being 20 amps in the worst type of conditions.

If we use two of these cables in parallel, we will have a total curent carrying capacity of 40 amps. As the main job of the circuit protective device is to protect the cable, this situation will be fine because the 30/32 amp device is less than the capacity of the cables in parallel (Figure 4.29).

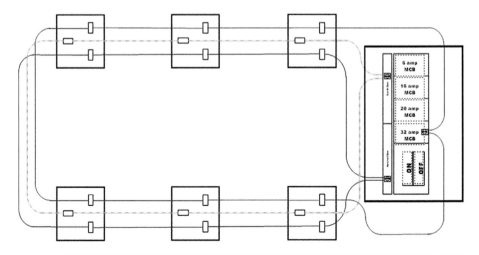

Figure 4.29 Ring circuit

Broken conductor in a ring circuit

One major problem with a ring circuit is that if one of the conductors were to be broken, or left disconnected for some reason, the protective device would not protect the cable from overload as it is rated at 30/32 amps and the single cable is only rated at 20 amps.

Apart from overloading of the conductors, the break or disconnection of a conductor (Figure 4.30) would result in the Z_s value of the circuit increasing. This, of course, in the event of an earth fault would have an effect on the operation of the circuit protection.

Interconnections

Apart from two or more rings being interconnected due to the cables getting muddled, situations can be found in ring circuits where a ring is wired within a ring.

This situation will not really present a danger, however it will make it very difficult for a ring circuit test to be carried out. Even if the correct ends of the ring are identified and connected together correctly, different values will be measures at various points of the ring. Also if one loop is broken, an end to end test of the ring will still show a continuous ring, and it will not be until further tests are carried out that the problem will be identified, and the broken loop or interconnection will be found (Figure 4.31).

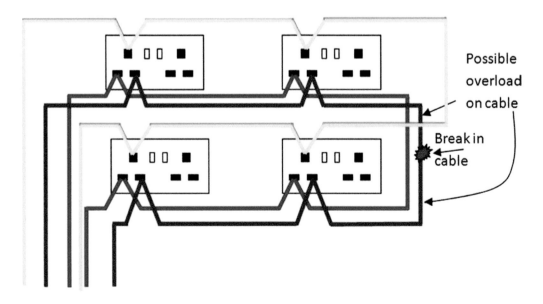

Possible overload on cable

Break in cable

Figure 4.30 Broken conductor

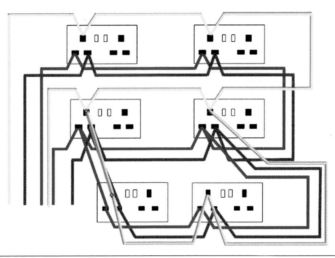

Figure 4.31 Interconnection

Polarity

Each socket outlet must be checked to ensure that the conductors are connected into the correct terminals. Clearly if they are not, serious danger could occur when appliances are plugged in.

Where the line and neutral are reversed polarity, the result would be that when a piece of equipment with a single pole operating switch is plugged in, it would switch off but would remain live. This could be particularly dangerous with something like a bedside lamp with an ES type lamp holder or an older type of electric fire with exposed elements.

If the line conductor and CPC were reversed, then in the case of any class 1 equipment becoming live, this could result in a fatal electric shock.

Performing the test

Ring final circuit test

Video footage is also available on the companion website for this book.

The test must be carried out using a low resistance ohm meter set on the lowest scale, typically 20Ω. The meter must always be checked for correct operation by joining the ends of the test leads together and checking for continuity, then test with the leads apart for open circuit. The leads must also be nulled/zeroed and the meter checked to ensure it is not damaged and that it is accurate.

Remember! This is a dead test and the circuit must be correctly isolated and locked off before working on it.

There are three steps to carrying out a ring final circuit test. However, before attempting the test the circuit must be completely isolated and the legs of the ring identified. The most obvious place to carry

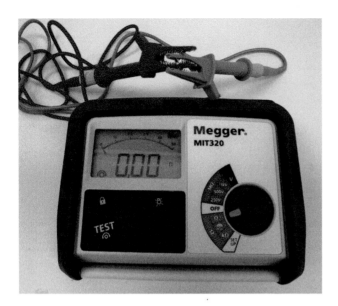

Figure 4.32 Set to ohms

out the test is at the consumer unit, although sometimes it is more convenient to carry the test out at a socket outlet.

Wherever the test is carried out the procedure remains the same. Check the instrument, null leads and set it to the lowest scale (Figure 4.32).

Step 1

Measure the resistance of each conductor end to end, this will show that each conductor is continuous and it will also provide you with a resistance value.

Test between the ends of the line conductor (Figure 4.33) and record the value on the schedule of test results in the correct column for R_1.

Test between the ends of the neutral conductor (Figure 4.34). This value should be the same as the line to line test as the conductors are the same size. Record this value as R_n.

Test between the ends of the CPC (Figure 4.35); this value will be higher than that of the live conductors if the CPC is smaller, for a 2.5mm^2 twin and earth cable with a 1.5mm^2 CPC the resistance will be 1.67 times greater than the resistance of the line or neutral conductor. If the CPC is the same size as the live conductors its resistance should be the same. Record this value as R_2.

Step 2

Link L1 to N2 and L2 to N1 (Figure 4.36) (in other words cross connect the opposite live ends of the ring).

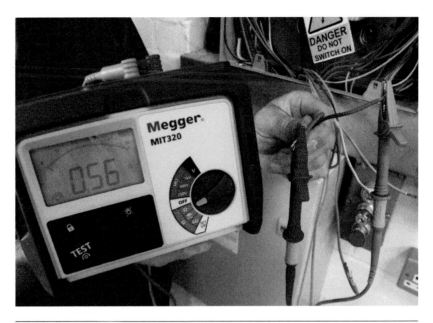

Figure 4.33 Test each end of the line conductor

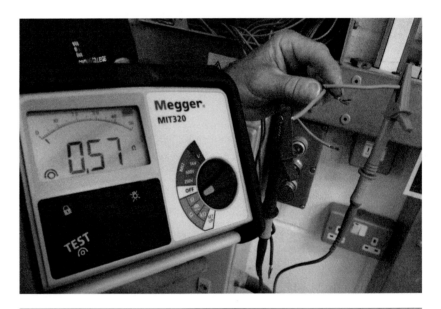

Figure 4.34 Test each end of the neutral conductor

Test between L and N at each socket outlet (make sure the socket switch is in the *on* position). At each socket outlet the resistance should be the same within 0.05Ω, unless of course the socket which is being tested is a spur, in which case the resistance will be higher (Figure 4.37).

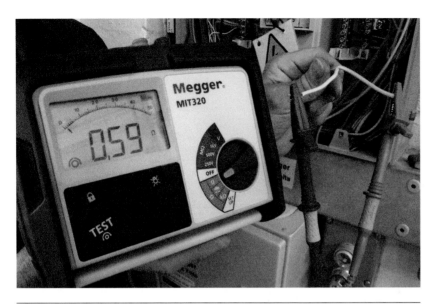

Figure 4.35 Test each end of CPC

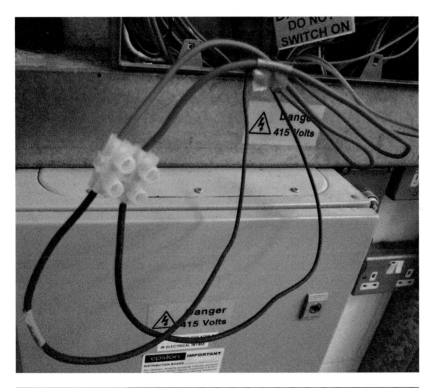

Figure 4.36 Cross connect live ends

Figure 4.37 Test line to neutral

Step 3

Link L1 to E2 and L2 to E1 (Figure 4.38) (cross connect the opposite ends of the L and E ends of the ring).

Test between L and E at each socket outlet (Figure 4.39). On circuits where the CPC is a smaller size than the line conductor, the resistance will be higher than the line to N resistance value and will increase slightly as you move towards the centre of the ring. On circuits where all of the conductors are the same cross-sectional area, the resistance value should be the same as the value measured between the L and N, again within 0.5Ω.

The highest measured value between line and earth should be recorded as the $R_1 + R_2$ value for the circuit.

Notes

1 If with the ends of the ring connected (L1/N2 and L2/N1) a substantially different value is measured at each socket outlet, then it is probably because it is not the opposite ends of the ring which are connected. Just swop them over and try again, this usually sorts the problem out.

2 In a twin and earth cable the resistance of the CPC will usually be 1.67 times that of the live conductors as it has a smaller cross sectional area.

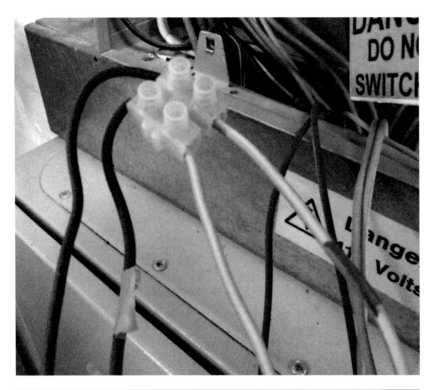

Figure 4.38 Cross connect line to earth

Figure 4.39 Test line to earth at each socket

3 When the live and CPC conductors are not the same size a higher resistance will be measured between line and CPC than line and neutral. It will also alter slightly as you move towards the centre of the ring. The resistance will be lower near the joined ends and will increase as you move away, the socket at the centre of the ring will have the same resistance as the joined ends.

4 Where circuits are contained in steel enclosures, parallel paths may be present. This would result in much lower $R_1 + R_2$ values than expected.

Example 4

Let's look at a ring circuit which is wired in 70°C thermoplastic twin and earth cable with 2.5mm² live conductors and a 1.5mm² CPC. The cable is 47 metres long end to end.

If we look at Table 9A in the *On-site Guide* we can see that a 2.5mm² copper conductor has a resistance of 7.41mΩ per metre, and that a 1.5mm² conductor has a resistance of 12.1mΩ per metre.

Step 1

Step 1 is to measure the resistance of each conductor end to end:

$$\text{L-L} \quad \frac{7.41 \times 47}{1000} = 0.34\Omega$$

$$\text{N-N} \quad \frac{7.41 \times 47}{1000} = 0.34\Omega$$

$$\text{E-E} \quad \frac{12.1 \times 47}{1000} = 0.56\Omega$$

These are the values which we could expect providing all of the conductors are connected properly and form a complete loop.

Step 2

Now we must join the line conductor from leg 1 to the N of leg 2, and the line conductor from leg 2 to the N of leg 1 (Figure 4.40).

Measure between L and N at each socket outlet and the resistance value should be half of resistance of one of the live conductor end to end measurement. The value we expect would be:

$$\frac{0.34}{2} = 0.17\Omega$$

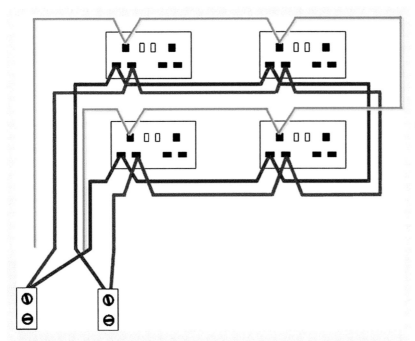

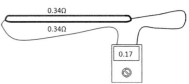

Figure 4.41 Conductors in series

Figure 4.42 Conductors in parallel

Figure 4.40 Cross connect live conductors

This is because that by interconnecting the ends of the ring we have halved the length and doubled the cross sectional area.

If we have one conductor which has a resistance of 0.34Ω and join another conductor with the same resistance to it, the length will now have a resistance of 0.68Ω (Figure 4.41)

Now if we join the other ends together we have halved the length of the conductor. This means that the resistance must be halved:

$$\frac{0.68}{2} = 0.34\Omega$$

And of course if we measure the loop from end to end we are also measuring double the cross sectional area which of course will result in the value halving again (Figure 4.42):

$$\frac{0.34}{2} = 0.17\Omega$$

The simple calculation to find the expected $R_1 + R_n$ value at each socket outlet is:

$$\frac{0.34 + 0.34}{4} = 0.17\Omega$$

Step 3

The same principle applies when we need to calculate the $R_1 + R_2$ value at each socket; the only difference is that in this example the cross sectional area of the conductors are different as we have a 2.5mm² line conductor and a 1.5mm² CPC. The same calculation can be used:

$$\frac{0.34 + 0.56}{4} = 0.225\Omega$$

This will be the value of $R_1 + R_2$ for the circuit. Always remember though that if there is a spur on the circuit that the $R_1 + R_2$ will be higher at the spur.

Truth table

Some training centres use what they call a truth table, where you make a note of the values as you record them. They are then compared with measured values as the test progresses, which will give you an early warning if the test begins to show up values which are different than those expected. As an example let's complete a truth test for the following ring circuit.

A ring final circuit is wired in 4mm² twin and earth 70°C thermoplastic cable which has a 1.5mm² CPC. The circuit is 63m long with no spurs.

First measure the line conductor end to end. This is recorded as R_1; the measured value should be $\dfrac{4.61 \times 63}{1000} = 0.29\Omega$.

If we now make a note of the value, it is also showing us the expected end to end resistance value of the neutral conductor R_n. This should be +/− 0.05Ω the value of R_1.

The next part of the test is to cross connect the line and N conductors of the ring. If we look at the truth table we will know that the expected measured value between L-N at each point is going to be half of the measured R_1 or R_n value which would be 0.15Ω or if we preferred we could add R_1 and R_n and divide the value by 4, this of course would give us the same value of around 0.15Ω. I find that it is much easier just to halve one of the values as I can do that in my head, providing the test between L-N is within 0.05Ω of my calculated value all is deemed to be ok.

Next we need to measure CPC end to end (R_2) and we would expect the value to be $\dfrac{12.1 \times 63}{1000} = 0.76\Omega$ or if we wanted to we could

Table 4.3 Truth table

R_1	R_n	R_2	R_1R_n	R_1R_2
0.29Ω	0.29Ω	0.77Ω	0.15Ω	0.26Ω

just multiply the R_1 value by the ratio of the difference between a 4mm² conductor and a 1.5mm² conductor. This can be found by dividing 4 by 1.5.

$$\frac{4}{1.5} = 2.66$$

$$2.66 \times 0.29 = 0.77Ω$$

Now we can enter the value into our truth table – 0.77. Next we add R_1 and R_2:

$$0.29 + 0.77 = 1.06$$

From our truth table we can see that the expected measured value between L and E when cross connected should be $\frac{1.06}{4} = 0.26Ω$

again this could be just calculated in your head.

A truth table for this ring circuit would look like Table 4.3.

Insulation resistance test

This test can be carried out on a complete installation, a group of circuits or on a single circuit, whichever is most suitable for the installation being tested. This test is carried out to find out if there is likely to be any leakage of current through insulated parts of the installation.

A leakage could be between live parts or live parts to earth; of course earth could be a metallic part of a piece of class 1 equipment, or any extraneous conductive part.

As we know voltage can be related to pressure and the pressure of our low voltage single phase supply is 230 volts a.c. When we carry out an insulation resistance test on a standard installation we must use a test voltage of 500 volts d.c. This is more than double the normal voltage and what we are doing can be related to a pressure test to identify any leaks.

Low insulation resistance

Cable insulation can deteriorate due to the age and type of cable insulation. The older type of rubber insulated cables tend to get very brittle where the outer mechanical protection has been removed leaving the insulated conductors exposed and this often causes a low installation resistance. Other causes due to cables could be where cables are crushed under floor boards, clipped on the edge or even worn very thin due to being pulled through holes in joists where other cables are present.

A low insulation resistance will indicate this type of damage. It will also often be found where the building has been left unused for a long period of time, particularly around cold and damp times of the year as dampness can creep into the accessories. In buildings which have just been plastered such as new build or renovations it is not unusual to get low readings due to the moisture in the walls causing dampness in switches or socket outlets. And of course it is very important that all of these conditions are identified.

Low resistance values can also be recorded where there are long cable runs or even where circuits are measured together in parallel, due to the amount of insulation being tested. (The longer the cables or the greater the number of circuits, in theory there is more insulation for the test current to leak through.)

The instrument used is a called an *insulation resistance tester*, it is also often referred to as a *high resistance ohm meter*. The tester has to meet certain requirements which are set by the Health and Safety Executive and these are that it must be capable of maintaining a test voltage of 500 volts d.c. and a current of 1mA when applied to a resistance of 1MΩ.

Table 4.4 shows the minimum acceptable insulation resistance values for circuits operating at various voltages, as described in BS 7671 (Table 61) – remember that this is a pressure test. This means that in most cases the test voltage should be greater than the operating voltage of the circuit, although under some circumstances this may be reduced due to the type of circuit or equipment connected to it.

Table 4.4 Minimum acceptable resistance values

	Circuits from 0v to 50v a.c.	Circuits from 50v to 500v a.c.	Circuits from 500v to 1000v a.c.
Required test voltage	250v d.c.	500v d.c.	1000v d.c.
Minimum acceptable insulation resistance	0.5 MΩ	1 MΩ	1 MΩ

Domestic installation

During an initial verification it is important that all conductors are tested, as under some circumstances this is impossible to do once the work is fully completed.

Always remember that inspecting testing should be carried out during erection and on completion (Regulation 641.1). One of the reasons for this is that once some pieces of equipment are connected, such as fluorescent lamps or transformers, it is impossible to test between live conductors as the load will have the effect of giving a poor reading.

Another good reason for carrying our insulation tests on cables is to check that they are not damaged before they are covered by building materials. This is particularly important on jobs that are slow to progress, and where parts of the installation are left for weeks on end without being worked on by the electrician. Under these circumstances the cables are pretty vulnerable and can easily be damaged – it is better that you find out while you have an opportunity to carry out the repair without taking floors up or ceilings down.

Testing the whole installation

Video footage is also available on the companion website for this book.

During an initial verification, particularly a domestic one, it is often easier to carry out the insulation test at the ends of the meter tails before they are connected to the supply. If the testing has been ongoing during the installation you will have a reasonable idea of the sort of values which you will be expecting to measure – hopefully very high!

It is very important to remember that if the board has any RCBOs to BS EN 61009, there is a very high possibility that very low readings will be measured. This is due to the construction of RCBOs and is not an indication of a fault. It is always worth just doing a test from the tails at 250 volts, but if a low reading is shown then you will have to disconnect each of the circuits which are protected by RCBOs and test them individually. All of the remaining circuits can be tested together through the tails. Where an RCD main switch to BS EN 61008 is used it is unlikely that it will cause any problems with the readings but again a first test of 250 volts is always a good idea.

As with any kind of dead test, ensure the system is dead. In this case it will be as the tails are not connected.

First check that the insulation resistance tester is working by testing it with the leads joined and then apart; with the leads together the reading should be 0.00Ω (Figure 4.43) and when apart it should be over range (Figure 4.44).

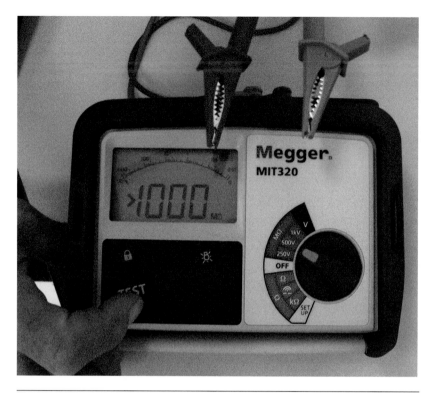

Figure 4.43 Leads apart

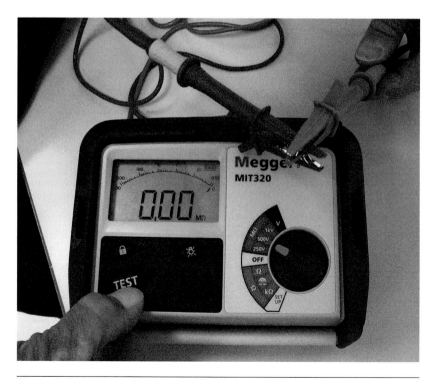

Figure 4.44 Leads together

Before proceeding with the test:

- Ensure the tester is accurate and that the test leads are compliant with GS38.
- Set the tester to 500 volts d.c. ready for the test.
- Check that the protective devices are in place and are switched on.
- Ensure that the main switch of the consumer unit is on.
- Remove all lamps from fittings where accessible.
- Providing testing has been ongoing, and lamps are not accessible, ensure that the switch controlling the luminaires is open (off); this also applies to luminaires with control gear such as fluorescent fittings. When testing installations which have not been subjected to testing during the installation, the luminaires must be disconnected and all of the conductors must be tested (note 1). The same applies where extra low voltage transformers are fitted.
- Where dimmer switches are fitted it is important that they are either removed and the switch wires joined together, or the switch can be linked across or bypassed (note 2).
- Any accessories which have neon indicator lamps fitted are switched off (note 3).
- Passive infrared detectors (PIRs) are removed, bypassed or linked out (note 4).
- All fixed equipment such as cookers, immersion heaters, heating circuits for boilers, TV amplifiers, etc. are all isolated.
- Shaver sockets are disconnected or isolated (note 5).
- All items of portable equipment are unplugged.

Notes

1 The control equipment inside discharge lamps will cause very low insulation resistance readings. It is quite acceptable to isolate the fitting by turning off the switch, this is far more desirable than disconnecting the fitting. After the test between live conductors has been carried out the control switch for the luminaire should be closed before carrying out the test between live conductors and the CPC. This is to ensure that all of the conductors are tested for insulation resistance to earth.

2 Most dimmer switches have electronic components in them and these could be damaged if 500V were to be applied to them. It is important that wherever possible dimmer switches are removed and the line and switch return are joined together for the test.

3 The test will also return a very low reading if neon indicator lamps are left in the circuit as they will be recognised as a load. All that is required is that the accessory is switched off.

4 Passive infrared detectors and light sensitive switches will also give very low readings and will be damaged by the test voltage. Either disconnect them, link them out or just test between live conductors and earth only.

5 Shaver sockets can also cause a problem; the best way to deal with them is to disconnect either the line or the neutral.

6 Wherever you are unsure of what is connected to the circuit it is always better to test at 250 volts first; if the reading is low do not proceed with the test until the reason for the low reading is identified.

Step 1

Set the insulation resistance tester to the required test voltage (Figure 4.45); for most low voltage circuits this will be 500 volts d.c.

Some instruments have settings for MΩ and some are self-ranging. Where they require setting, 200MΩ or above is the most suitable setting to use.

Step 2

Always check that the tester is working correctly by testing with the leads apart (Figure 4.43). The reading shown should be the highest that the tester should measure.

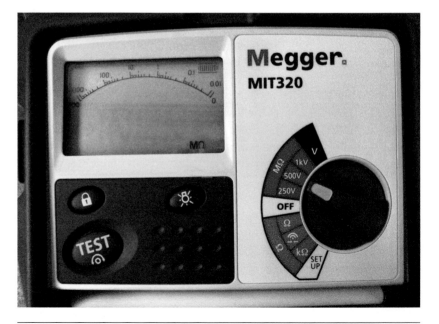

Figure 4.45 Instruments set

Step 3

Now join the leads together and operate the tester again. This should produce a reading of 0.00MΩ which is the lowest that the tester will read (Figure 4.44). This proves that the tester is working and that the leads are not broken.

Step 4

When testing a complete installation there are options as to how the test is carried out – you can test from the tails if it is a new installation which has not been connected, or you can test from the dead side of the main switch if the installation has been connected. An example would be a rewire.

Ensure that all of the circuit breakers are in the on position or if the installation is protected by fuses make sure that they are all in place.

Where the test is being carried out at the main switch, make sure that it is in the off position and locked off.

Test between live conductors (Figure 4.46) and operate any two way switching. This is to ensure that the strappers of the circuit are tested and it will also ensure that the switch returns have been correctly identified and connected (no neutrals in the switches).

Step 5

Test between live conductors and earth (Figure 4.47). This can be by joining together the live conductors, or if testing at the main switch

Single phase insulation test

Video footage is also available on the companion website for this book.

Figure 4.46 Test between live conductors

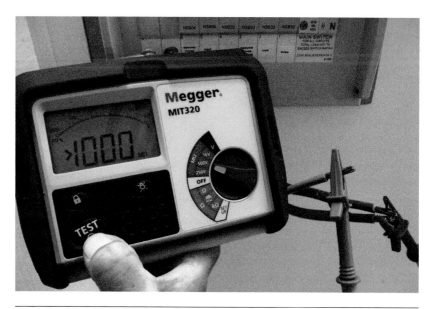

Figure 4.47 Join live conductors and test to earth

it can be carried out between each conductor individually. Again it is important to operate any two way switches.

As the test is being carried out the measured values should be entered on to the schedule of test results. Table 61A of BS 7671 shows the acceptable insulation resistance as being 1MΩ, this is for a single circuit or a complete installation. Guidance Note 3 recommends that any circuit giving a value of 2MΩ or less must be investigated; this is because a value as low as 2MΩ may indicate a latent defect which could develop into a major problem at a later date.

Although these values will comply with the requirements of BS 7671, they are still very low values for the majority of circuits.

Consideration must be given as to why the circuit insulation resistance is low; where the circuit is new it is very unlikely that a reading of 2MΩ would be acceptable. Even on an older existing circuit there would need to be a good reason why the value is so low, and it would need to be monitored to check that it was not getting worse.

Where the whole installation is being tested in one go and a low reading is measured, it is a good idea to test each circuit individually, as this will identify whether it is one circuit causing the problem or an accumulation of the insulation resistance of circuits being measured in parallel. In theory the more circuits in parallel the lower the reading will be, although in practice it really should not make much difference providing the circuits are in good condition.

On some occasions a low value may be acceptable, often where a building has been empty for a long period of time, particularly in the winter, or perhaps some of the installation is outside or underground cables have been installed. Whatever the case may be, it is important that you have some idea of why low readings, although acceptable, are being recorded as it may be indicating a problem for the future. Of course in the case of an unused installation the insulation resistance will probably rise after a period of use.

Example 5

An installation consisting of six circuits is tested as a whole, the insulation resistance between live conductors and earth is 1.9MΩ.

The installation is split and each circuit is now tested individually. All of the circuits are found to be between 100 and 200MΩ apart from one. On investigation it is found that the circuit causing the low reading is a mineral insulated cable supplying a lamp post in the garden, so it could be that the connection to the lamp is damp or has condensation in it. In this case it is easily rectified; in cases where the problem is not so easily found the circuit should be monitored, as often if a circuit which is wired with mineral insulated cable is left switched on with a load connected it will dry out naturally.

In all cases where insulation resistance tests are carried out it is important that the results are thought about and not just recorded; an element of careful thought and sensible judgement has to be part of the testing process.

Where a complete installation is tested from the tails or main switch and an acceptable value is measured, it is permissible to enter the same value for all of the circuits on the schedule of test results.

Testing of individual circuits

On existing installations or even some new installations it may be necessary to test each circuit individually, particularly where the complete installation cannot be isolated. The same safety precautions must be taken.

Always ensure that the circuit to be tested has been safely isolated by removing the fuse or turning off the circuit breaker; always follow the isolation procedure and make sure that the circuit is locked off, and that you are the only person who can switch it back on.

It will be necessary to disconnect the neutral of the circuit being tested; this is because all of the neutrals are connected into a common terminal. On installations which have to remain live, any neutral connected to the neutral bar will be also connected to the star point of the supply transformer, which in turn is connected to earth. This of course will produce a very low reading between neutral and earth.

Never disconnect the CPC of the circuit from the main earth terminal when carrying out this test; this is because with the CPC disconnected the test will only indicate a fault between the conductors of the cable being tested. It may be that one of the live conductors is touching an extraneous or exposed conductive part within the installation. If the CPC is disconnected the fault will not show up when a live conductor to earth test is carried out; however, if the CPC is connected to the main earthing terminal the fault will show.

Check that all equipment which may be vulnerable to testing, or any equipment which may produce a low reading is disconnected or isolated.

Remember if in doubt about what may be connected, test at 250 volts first.

Carry out the checks on the tester and leads to ensure its correct operation.

As previously mentioned, if the main switch is off then the tests can be carried out without disconnecting the conductors. They will only need to be disconnected if the measured value is low, it may be that the low reading is not on the circuit which is being tested. This is because all of the neutrals and all of the CPCs are connected into common terminals.

Test between live conductors (Figure 4.48) and then live conductors and earth (Figure 4.49).

When testing between live conductors and earth, the live conductors can be joined together and then tested to earth or they can be tested separately, whichever is easier.

Where for some reason a piece of equipment connected to the installation cannot be isolated from the circuit being tested, do not carry out the test between live conductors – only test between live conductors and earth. This will avoid damage to the equipment due to the test voltage; where this type of test is used, only complete the L-E box on the schedule of test results and enter N/A in the L-L box.

This test method should only be carried out on individual circuits and not on a whole installation, it is important that as much of the installation is tested as possible.

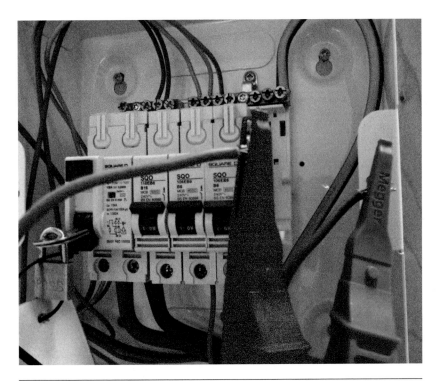

Figure 4.48 Test between live conductors

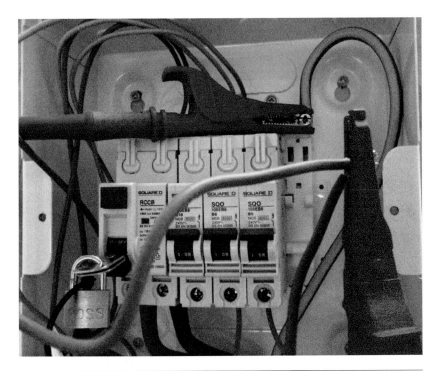

Figure 4.49 Live conductors tested to earth

Surge protection

Where a circuit has surge protection an insulation test can give very low readings; this will of course look like there is a fault when in fact it is really just that surge protection has been fitted. Where possible it is better to disconnect the surge protection; however where this is not an option, it is perfectly acceptable to test at 250v d.c. and record the values obtained. Surge protection devices are designed to leak to earth at voltages above 250 volts and of course this will then show up as a bad reading.

Three phase test

Video footage is also available on the companion website for this book.

Insulation resistance testing of a three phase installation

When carrying out an insulation resistance test on a three phase installation all of the same precautions and procedures apply as for a single phase installation or circuit.

Where the test is for the whole installation the test can be carried out on the isolated side of the main switch. Safe isolation must be carried out before beginning the test and all of the protective devices must be in the on position.

Step 1

Set the tester to 500v d.c. (Figure 4.50).

Step 2

Test between all line conductors (Figure 4.51).

Step 3

Test between all line conductors and neutral (Figure 4.52); the line conductors can be joined together for this part of the test.

Step 4

Test between all live conductors to earth; the line and N conductors can be linked for this part of the test (Figure 4.53).

When testing a three phase installation the minimum resistance values are the same as for single phase installations. The minimum acceptable is 1MΩ although any values below 2MΩ must be investigated.

Step 5

Remove any links which have been used to simplify the testing process.

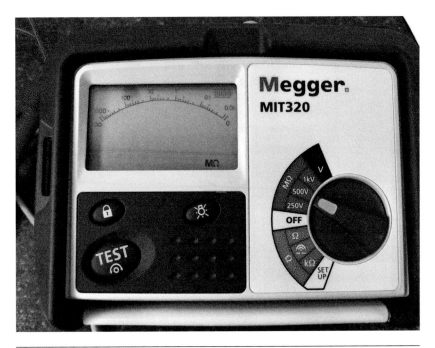

Figure 4.50 Test the set

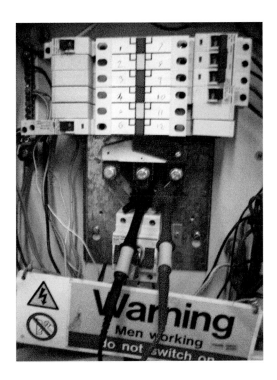

Figure 4.51a Test L2 to L3

Figure 4.51b Test L1 to L3

Figure 4.51c Test L1 to L2

Figure 4.52 Link L1, L2 and L3

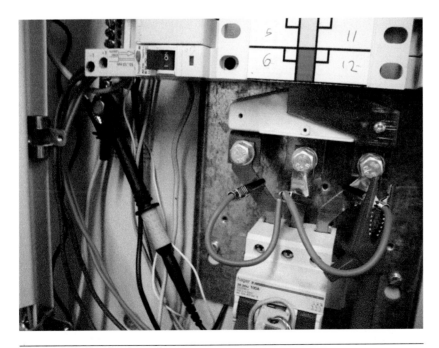

Figure 4.53 Test line to earth

Video footage is also available on the companion website for this book.

On occasions where it is only possible to test each circuit individually, it is still very important to ensure that the insulation resistance for the whole installation does not fall below the minimum acceptable value which of course is 1MΩ. Under these circumstances a calculation must be carried out.

Example 6

A consumer unit has 6 circuits which when measured for insulation resistance between live conductors and earth.

Circuit 1 is 140MΩ

Circuit 2 is 70MΩ

Circuit 3 is 10MΩ

Circuit 4 is 8MΩ

Circuit 5 is 200MΩ

Circuit 6 is 45MΩ

We now need to carry out a calculation to check that the insulation resistance value of the total installation:

The calculation is: $\dfrac{1}{R_1}+\dfrac{1}{R_2}+\dfrac{1}{R_3}+\dfrac{1}{R_4}+\dfrac{1}{R_5}+\dfrac{1}{R_6}=\dfrac{1}{R_t}=R$

Put in the values: $\dfrac{1}{140}+\dfrac{1}{70}+\dfrac{1}{10}+\dfrac{1}{8}+\dfrac{1}{200}+\dfrac{1}{45}=0.27(R_t)$

Now we must remember to complete the last stage which is:

$\dfrac{1}{R_t}$ or in figures $\dfrac{1}{0.27}=3.65MΩ$

These figures can be entered into a calculator as follows:

$140x^{-1}+70x^{-1}+10x^{-1}+8x^{-1}+200x^{-1}+45x^{-1}=x^{-1}=3.65$

This value is greater than 1MΩ and is satisfactory.

When completing a schedule of test results, it is important that you record which insulation tests have been carried out. Where it is not possible to carry out tests between any conductors, a note of this should be made in the remarks column of the schedule of test results. This is of course providing the test is being carried out during a periodic inspection and test; where the test is carried out during an initial verification, the test must be carried out between all conductors.

Polarity test

This is a test which is carried out to ensure that:

- All protective devices are connected into the line conductor of the circuit which they are protecting.
- Single pole switches are connected in the line conductor.
- ES lampholders have the centre pin connected to the switch return (except for E14 and E27 lampholders complying with BS EN 60238) Regulation 643.6. These type of lampholders are all insulated.
- All accessories such as fused connection units, cooker outlets and the like are connected correctly.

In many instances this test can be carried out at the same time as the $R_1 + R_2$ or CPC continuity test. The instrument required is a low resistance ohm meter.

This is a dead test and the installation must be isolated and locked off before starting the test.

Polarity test on a radial circuit such as a cooker or immersion heater circuit

Step 1

At the origin of the circuit, link the line and CPC using a short lead with crocodile clips at each end or simply just connect the CPC to the line conductor by putting one or the other in the same terminal or connector (Figure 4.54). It is also a good time to check that the conductors are identified correctly.

Step 2

At the furthest point of the circuit remove the cooker or immersion heater switch, visually check that the conductors are connected into the correct terminals and that they are correctly identified.

Step 3

If the test is also for $R_1 + R_2$ remember to null the test leads.

Connect the test leads to the line and earth at the accessory and carry out the test (Figure 4.55). The reading should be very low and it will also be the $R_1 + R_2$ for the circuit.

Step 4

Remove the lead from the consumer unit and test at the accessory again; this time the circuit should have a high reading (Figure 4.56). This will prove that the correct circuit is being tested.

Figure 4.54 Line and CPC connected

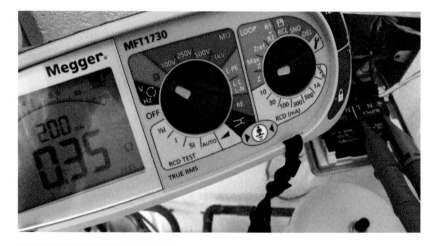

Figure 4.55 Test between line and earth R$_1$ + R$_2$

Polarity test on a lighting circuit

Video footage is also available on the companion website for this book.

Polarity test on a lighting circuit

Step 1

At the origin of the circuit join the line and CPC of the circuit, either with a lead or simply connect together (Figure 4.57).

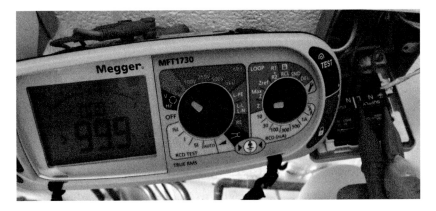

Figure 4.56 High reading

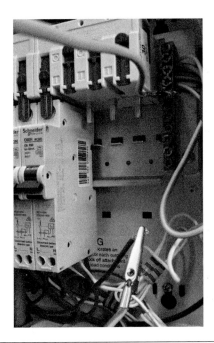

Figure 4.57 Line linked to earth terminal

Step 2

At the light fitting place the probes of the leads onto the earth terminal and the switched line (Figure 4.58).

Step 3

Now close the switch controlling the light and the instrument should give a low reading (Figure 4.59) (this will also be the $R_1 + R_2$ value if when the point tested is the furthest from the consumer unit). Now open the switch and the reading should be very high (Figure 4.60), in fact the tester should be over range.

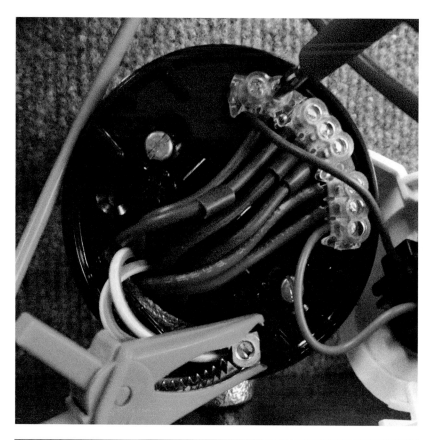

Figure 4.58 Test between earth and switched line

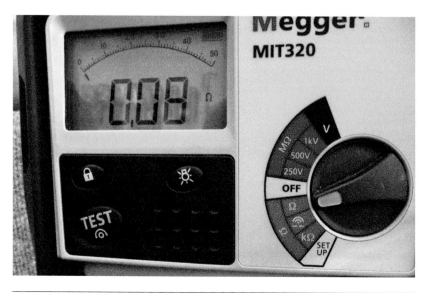

Figure 4.59 Low reading

Figure 4.60 High reading

This test can be carried out by linking the earth and switched line at the lighting point, and testing from the consumer unit the end result will be the same.

On occasion it may be difficult to access the light point and on these occasions the test can be carried out at the light switch.

Step 1
Place a link between the line and CPC of the circuit at the consumer unit (Figure 4.61).

Step 2
At the switch, place the probes on the earth terminal and the switched return terminal (Figure 4.62).

Close the switch and a low resistance value should be shown on the tester (Figure 4.63).

Step 3
Open the switch and the instrument should now read over range as the circuit will be open circuit (Figure 4.64).

Figure 4.61 Line and earth terminal linked

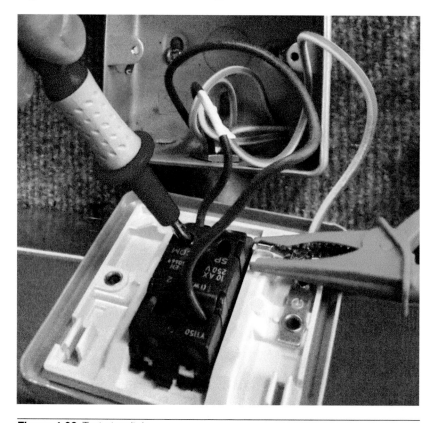

Figure 4.62 Test at switch

Figure 4.63 Low resistance measure

Figure 4.64 High reading

Live polarity

This test is usually carried out at the origin of the supply before it is energised to ensure that the supply being delivered to the installation is the correct polarity. Although this is a test it does not really count as part of the sequence of tests as it is just a check that the supply is correct.

The instrument used is an approved voltage indicator or test lamp which complies with HSE document GS38. It is also acceptable to use an earth fault loop impedance tester as they also indicate polarity.

Step 1

Test between line and neutral at the main switch; the device should indicate a live supply (Figure 4.65).

Step 2

Test between line and earth bar (Figure 4.66); the device should indicate a live supply.

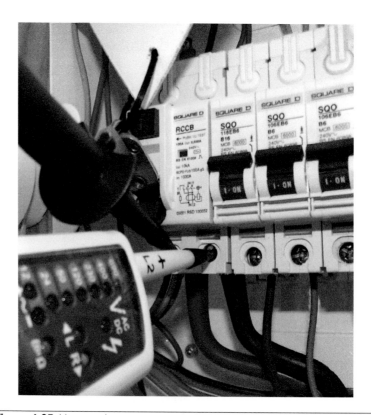

Figure 4.65 Live supply

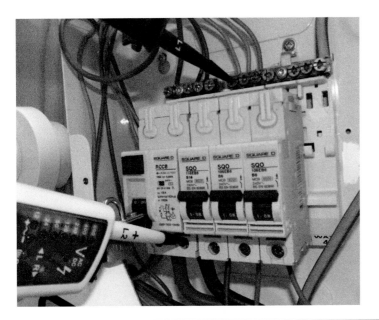

Figure 4.66 Live supply

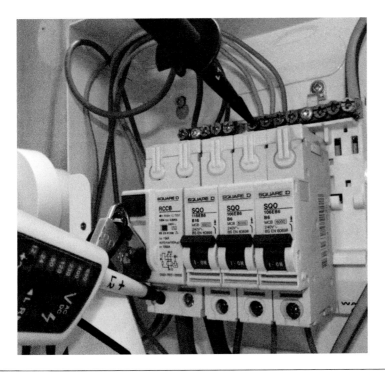

Figure 4.67 No reading

Step 3

Test between neutral bar and earth bar (Figure 4.67); the device should indicate no supply.

CHAPTER 5

Earth electrode testing

There are two methods which can be used to test an earth electrode and on most installations it is perfectly acceptable to use an earth fault loop impedance tester. An earth electrode resistance tester would normally be used for special installations, or where the electrode is for a generation system and low accurate readings are required.

Most electricians will only use earth electrodes on TT systems where a reasonably high resistance value will be expected.

Where lower and more accurate values of earth electrode resistances are required, an earth electrode resistance tester should be used.

Earth electrode testing

Video footage is also available on the companion website for this book.

Measurement using an earth electrode tester

This test requires the use of three electrodes, the electrode under test which is referred to as E, a current electrode and a potential electrode. The current electrode and potential electrode are really spikes as shown in Figure 5.1.

Performing the test

Step 1

The electrode under test (E) should be driven into the ground in the position at which it is going to be used (Figure 5.2). The length of the electrode should be measured before it is installed as it is helpful to know how much of it is in the ground.

Step 2

Place the current spike (C) into the ground a minimum of 10 times. So if E is buried to a depth of 3m, this means that C must be a minimum of 30m away from E.

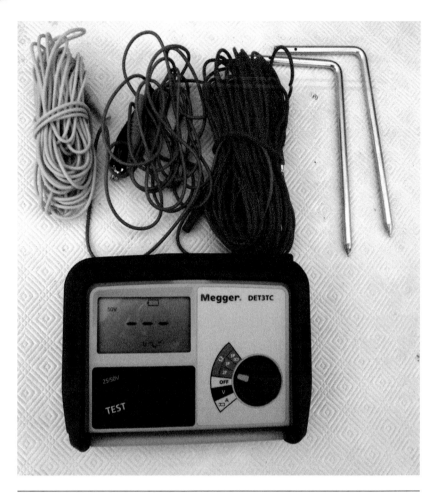

Figure 5.1 Earth electrode tester

Step 3

The potential electrode (P) must now be placed approximately in the centre between electrodes E and C.

Step 4

The leads of the test instrument should now be connected to the electrodes. Some testers have 4 leads which are identified as $C_1, P_1,$ C_2 and P_2. Others have only 3 leads which will be clearly identified, usually as E, C and P.

For a 4 lead tester the C_1 and P_1 leads should be connected to E. C_2 and P_2 are connected to their respective electrodes. For a three lead tester the connection is obvious with E, P and C connected to their electrodes.

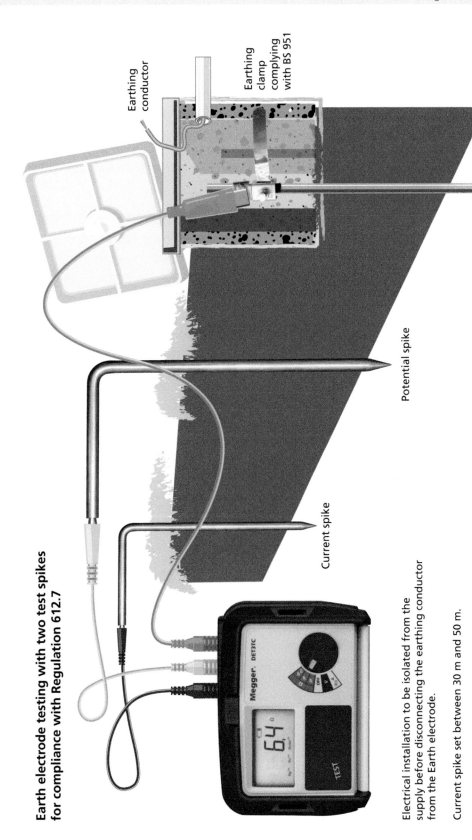

Earth electrode testing with two test spikes for compliance with Regulation 612.7

Earthing conductor

Earthing clamp complying with BS 951

Installation earth electrode under test

Potential spike

Current spike

Electrical installation to be isolated from the supply before disconnecting the earthing conductor from the Earth electrode.

Current spike set between 30 m and 50 m.

Potential spike set in line and equidistant between electrode under test and the current spike. For more accurate measurements set the potential spike at 62% of the distance between electrode under test and the current spike.

Reconnect earthing conductor to Earth electrode before re-energized the installation.

Figure 5.2 Electrodes in the ground

Step 5

Measure the resistance and make a note of it (let's say it is 79Ω).

Step 6

Move the middle electrode P 10 per cent (3m) closer to C.

Step 7

Measure the resistance and make a note of the resistance (let's say it is 85Ω).

Step 8

Now move spike P 10 per cent (3m) from the centre closer to E.

Step 9

Measure and make a note of the resistance (let's say 80Ω).

A calculation must now be carried out to find the percentage deviation of the resistance value. This should be no more than 5 per cent.

Example 1

Add the three values together and calculate the average value.

The total value of the three reading is:

$$79 + 85 + 80 = 244$$

The average is found by:

$$\frac{244}{3} = 81.33\Omega$$

Now we need to find the difference between the average value and the other values. The highest value is the one which we will use to complete our calculation.

$$81.33 - 79 = 2.33$$
$$81.33 - 80 = 1.33$$
$$85 - 81.33 = 3.67$$

The highest value is 3.67Ω.

The percentage value to the average value must now be found:

$$\frac{3.67 \times 100}{81.33} = 4.51\Omega$$

This value is known as the percentage deviation.

Guidance Note 3 tells us that the accuracy of this measurement is typically 1.2 times the percentage deviation. Therefore to ensure that we use the correct value we must multiply the percentage deviation by 1.2.

$$4.51 \times 1.2 = 5.41\Omega$$

As this value is greater than 5 per cent of the average value (5 per cent of 81.33 = 4.06) it is not advisable to accept it. A deviation of greater than 5 per cent is deemed to be inaccurate. To overcome this, the distance between electrode E and spike C must be increased and the procedure and calculations must be repeated using the new values. This may need to be carried out a few times until an accurate value is obtained.

In circumstances where a low enough resistance value cannot be obtained by a single electrode, additional electrodes can be added at a distance from electrode E equal to the depth of E.

Testing with an earth loop resistance tester

Z_e measurement test for a TT system

Video footage is also available on the companion website for this book.

This is the most common method which is used for measuring the resistance of earth electrodes in domestic situations, or for electrodes where low accurate values are not required.

Before you start always remember to gain permission to isolate the system before starting.

Step 1

Isolate and lock off the installation.

Step 2

Ensure that the earthing conductor is correctly connected to the earth electrode.

Step 3

Disconnect the earthing conductor from the main earthing terminal

Step 4

Connect one lead of the earth fault loop impedance tester to the disconnected earthing conductor and place the probe of the other lead onto the incoming line conductor at the main switch (Figure 5.3).

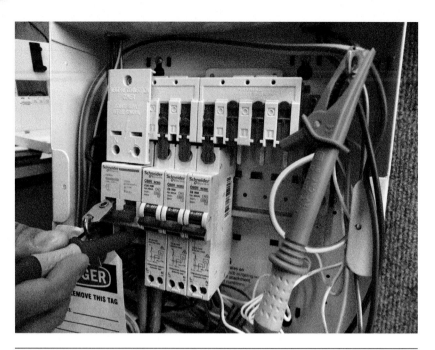

Figure 5.3 Tester connected

Step 5

Record the measured value

Step 6

Reconnect the earthing conductor and leave the installation safe.

If a three lead tester is to be used, read the tester instructions before attempting to carry out the test. On some testers it is necessary to join the N and E leads together and on others the leads need to be connected individually.

Table 41.5 from BS 7671 provides values for the maximum permissible loop impedance which will ensure correct operation of RCDs.

These values can also be found using the following calculation.

For TT systems the maximum touch voltage must not be allowed to rise above 50 volts under fault conditions; this is the value to be used for the calculation.

50v is the maximum voltage.

$I_{\Delta n}$ is the trip rating of the residual current device.

Z_e is the value of earth fault loop impedance (where measuring the resistance of the earth electrode this value is known as R_A).

If the rating of the RCD is 30 mA the calculation is:

$$\frac{50}{0.03} = 1667\Omega$$

For a 100 mA device the calculation would be:

$$\frac{50}{0.1} = 500\Omega$$

This calculation provides the maximum permissible value for the earth fault path which will cause the RCD to operate in the correct time. However an electrode providing this level of resistance would be seen as unacceptable. This is because the soil resistance may dry out which of course would result in the resistance increasing.

The acceptable value give in BS 7671 for a 500 or 30 mA RCD is 200Ω; any value above this would be seen as being unstable.

Where high values are measured it may be necessary to install an additional electrode or electrodes to bring the resistance value down to an acceptable level.

The maximum calculated values for earth electrodes are shown in Table 5.1.

For special locations where the maximum touch voltage is 25V the electrode resistance should be halved. Electrode tests should always be carried out under the worst possible conditions; these would be when the soil is *dry*.

This test is carried out using the same procedure as you would use to Measure Z_e.

Z_e is a measurement of the impedance (resistance) of the external earth fault path of the installation. In other words we are measuring the impedance of the supply transformer, the supply incoming line and the return path.

Table 5.1 Electrode resistance values

Operating current of the RCD ($I_{\Delta n}$)	Earth electrode resistance in Ω
30mA (0.03A)	1667
100mA (0.1A)	500
300mA (0.3A)	160
500mA (0.5A)	100

Earth fault path for a TT system

For a TT system the return path is through the earth electrode and mass of earth (Figures 5.4 and 5.5).

Earth fault path for a TN-S system

This system uses the metal sheath of the supply cable as the earth fault return path (Figures 5.6 and 5.7).

Earth fault path for a TN-C-S system

This system uses the protective earth and neutral conductor of the supply as the earth fault return path (Figures 5.8 and 5.9).

Figure 5.4 TT service head

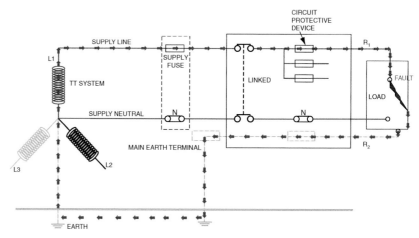

Figure 5.5 TT fault path

Figure 5.6 TN-S service head

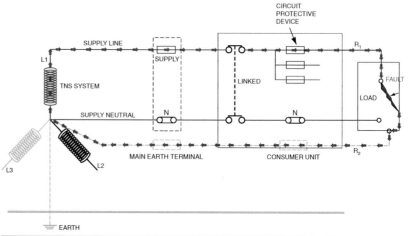

Figure 5.7 TN-S fault path

Figure 5.8 TN-C-S service head

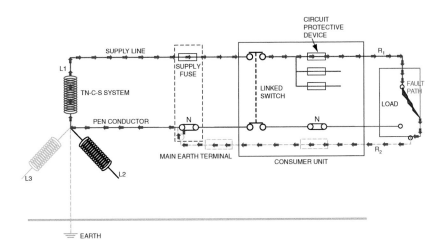

Figure 5.9 TN-C-S fault path

Performing a Z$_e$ test

Step 1

Isolate the supply and lock off (Figure 5.10).

Step 2

Disconnect the earthing conductor from the main earth bar (Figure 5.11).

Step 3

Use an earth loop impedance tester set to loop (Z$_e$) with leads to GS38 (Figure 5.12).

Figure 5.10 Isolated

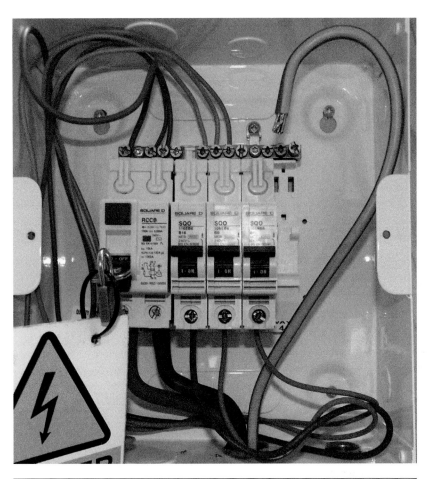

Figure 5.11 Disconnected earthing conductor

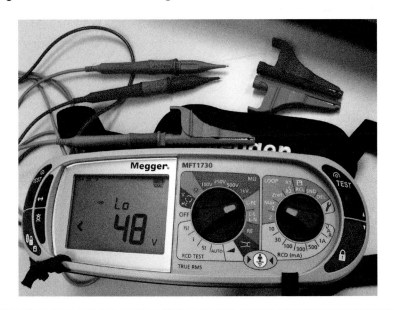

Figure 5.12 Test instrument

Step 4

If using a two lead tester connect one lead to incoming line and the other to the disconnected earthing conductor (Figure 5.13). If using a three lead tester it is important that the tester instructions are read first. Some three lead testers require the N lead to be connected to the neutral of the incoming supply and others require it to be connected to the earth along with the earth lead (Figure 5.14).

Step 5

The value measured will be the Z_e and should be recorded on the appropriate certificate (Figure 5.15).

Step 6

Reconnect the earthing conductor, replace all covers and re-energise the system.

It is very important before carrying out any live test that the instrument instructions are read and fully understood before attempting to carry out the test.

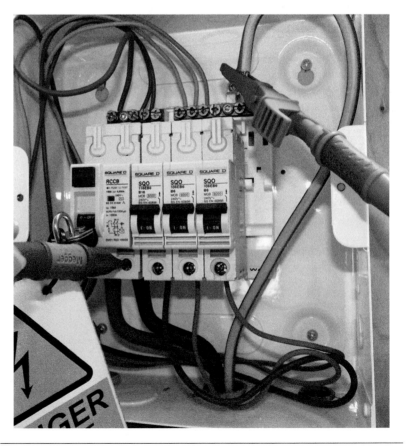

Figure 5.13 Line to earth

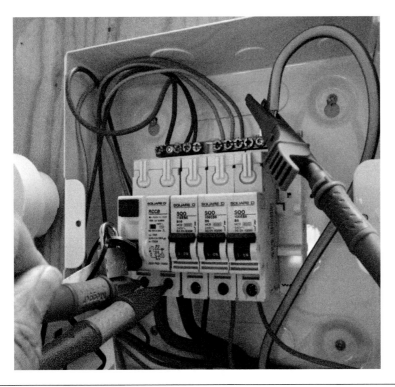

Figure 5.14 Line to neutral and earth

SUPPLY CHARACTERISTICS AND EARTHING ARRANGEMENTS			*(Tick boxes and enter details, as appropriate)*	
Earthing TN-C TN-S TN-C-S TT IT Other source of supply (to be detailed on attached schedules)	**Number and Type of Live Conductors** a.c. d.c. 1-Phase,2-Wire 2-wire 2-Phase,3-Wire 3-wire 3-Phase,3-Wire Other 3-Phase,4-Wire		**Nature of Supply Parameters** Nominal voltage, U/U_0 [1] V Nominal frequency, f [1] Hz Prospective fault current, I_{pf} [2] kA External loop impedance, Z_e [2] | *(Note: (1) by enquiry, (2) by enquiry or by measurement)*	**Supply Protective Device Characteristics** Type Rated Current A

Figure 5.15 Certificate

Circuit earth fault loop impedance Z_s

Z_s is the value of the impedance (resistance) of a final circuit and the supply fault loop path.

There are two methods which can be used to obtain the earth fault loop impedance of a circuit (Z_s).

One method is by direct measurement using an earth fault loop tester and the other is by calculation.

Video footage is also available on the companion website for this book.

Calculation

To calculate Z_s accurately, we must have already measured the value of Z_e and $R_1 + R_2$. It is important that the value of $R_1 + R_2$ is the highest value which has been measured on the circuit; this is normally at the end of the circuit. We also have to consider spurs from rings and radial circuits as these may not appear to be the furthest point but will often produce a higher reading due to the additional length of cable.

The calculation for Z_s is:

$$Z_s = Z_e + R_1 + R_2 \text{ (Figure 5.16)}$$

Let's say that the Z_e is 0.42 and that the $R_1 + R_2$ value is 0.68. All that is required is for us to add them together which of course in this example would give us a Z_s value of 1.1Ω.

Whichever method we use we need to verify that the end result is suitable and that it will ensure the correct operation of the device which is protecting the circuit.

Verification of Z_s values

The reason for having an accurate Z_s value is so that we can compare it with the maximum Z_s values provided in BS 7671; of course we must remember that they are the *maximum* values.

These maximum values are arrived at by the calculation:

$$Z_s \times I_a \leq U_0 \times C_{min} \text{ or transposed } Z_s = \frac{U_0 \times C_{min}}{I_a}$$

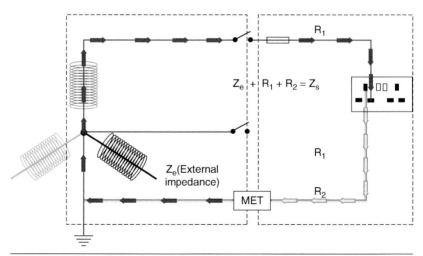

Figure 5.16 $R_1 + R_2$ path

In other words the $Z_s \times$ the current which will operate the protective device must not be greater than the supply voltage $\times$ the voltage factor of 0.95 which is used to take into account any voltage fluctuation of the supply. U_0 is always taken as 230 volts for a single phase supply.

As an example, if we look at a 6A type B circuit breaker we will see that the current causing automatic operation will be $5 \times 6 = 30A$

$$U_0 \times C_{min} \text{ will be } 230 \times 0.95 = 218.5 \text{ volts}$$

If we now look at the maximum Z_s values in BS 7671 we can see that the maximum Z_s value for a 6A type B circuit breaker is 7.28Ω.

Using our original calculation of:

$$Z_s \times I_a \leq U_0 \times C_{min} \text{ or with numbers } 7.28 \times 30 \leq 230 \times 0.95$$

we can see that:

$$7.28 \times 30 = 218.4v \text{ which is less than } 230 \times 0.95 = 218.5v$$

The values which we have calculated or measured will have been using the resistance values of the cable. When carrying out our measurements we will not have known the temperature of the circuit cables, or the temperature that the cables will reach when the circuit is under load.

This temperature must always be taken into account, as a rise in conductor temperature will result in a rise in conductor resistance.

Most of the cable which we use in electrical installations is rated to operate at 70°C. The resistance values given for our conductors in Tables B1 of Guidance Note 3, and Table I1 of the *On-site Guide* are for cables at a temperature of 20°C.

The resistance of the copper conductors will rise by 2 per cent for each 5°C rise in temperature. With that in mind if the $R_1 + R_2$ values are measured at 20°C, and the temperature of the conductor rises to its maximum of temperature of 70°C when it is carrying a load, we can see that the temperature rise of the conductors is 50°C.

As the resistance rises by 2 per cent for each 5°C rise in temperature then the resistance must rise by 20 per cent, as there is $10 \times 5°C$ in 50°C. If the rise is 2 per cent for each 5°C then the increase in resistance will be 10×2 per cent $= 20$ per cent.

To increase any value by 20 per cent we simply multiply it by 1.2. If we look in Table B3 of GN3 or I3 of the *On-site Guide* we can also see that 1.2 is the multiplier given for temperature correction.

There are different scenarios which we must consider.

Scenario 1

Where the $R_1 + R_2$ values for the circuit length have been calculated using the $r_1 + r_2$ values from Table I1 in the *On-site Guide* or Table B1 in Guidance Note 3.

Circuit is wired in 2.5mm^2 with a 1.5mm^2 CPC and is 21m long.

From the tables we can see that the value of $r_1 + r_2$ for this size copper cable is 19.51mΩ/m at 20°C.

21m of this cable will have a resistance $(R_1 + R_2)$ of

$$\frac{19.51 \times 21}{1000} = 0.4\Omega$$

This is of course at 20°C; we now have to calculate the resistance at which the conductor will be if it is operating at a temperature of 70°C.

To do this we must multiply the $R_1 + R_2$ value by the factor of 1.2.

$$0.4 \times 1.2 = 0.48\Omega$$

This will be the resistance which the conductor will reach at 70°C.

To calculate Z_s we must add 0.48 to the measured Z_e value. This value of Z_s can now be directly compared with the maximum values given for Z_s in Chapter 41 of BS 7671. Providing our value is equal to or less than the maximum value the circuit will be satisfactory.

Scenario 2

Where the length of the circuit is not known and the value of $R_1 + R_2$ has been measured using the $R_1 + R_2$ method, an accurate calculation can be obtained by using the multiplier values provided in Table B2 of GN3 and I2 of the *On-site Guide* as *divider values*. This is because the table is intended to be used to calculate the resistance value of a conductor when we know the temperature that it is going to operate at. In our case we are going to measure the temperature of the room, and calculate the resistance of the cable back to what it would be at 20°C.

Ambient temperature multipliers to be applied to $R_1 + R_2$ values are shown in Table 5.2.

As an example let's say that we have measured an $R_1 + R_2$ value of 0.84Ω and we have measured the ambient temperature at 25°C.

The calculation is:

$$\frac{R_1 + R_2}{temp\ factor} = the\ value\ of\ resistance\ at\ 20°C$$

$$\frac{0.84}{1.02} = 0.82\Omega \text{ is the resistance of the cable at 20°C}$$

Table 5.2 Ambient temperature multipliers

Expected or measured ambient temperature	Multiplier values
5°C	0.94
10°C	0.96
15°C	0.98
20°C	1.00
25°C	1.02

Having corrected the measured value to what it would be at 20°C the next step is to calculate what the resistance of the cable would be at its operating temperature.

This is where we use the 1.2 multiplier.

The resistance of the cable at its maximum operating temperature of 70°C:

$$0.82 \times 1.2 = 0.98\Omega$$

This value can now be added to the measured Z_e to provide a value of Z_s which can be compared directly to the maximum Z_s values provided in Chapter 41 of BS 7671.

Example 2

The Z_e of an installation is 0.32Ω. A circuit has been installed using twin and earth 70°C thermoplastic (pvc) cable. The room temperature is 25°C and the measured $R_1 + R_2$ value is 0.52Ω. The circuit is protected by a BS EN 60898 20A type C circuit breaker.

Correct the cable resistance to 20°C by using the factor from Table B1 or I2.

$$\frac{0.52}{1.02} = 0.51\Omega$$

Adjust this value to the conductor operating temperature by increasing it by 20 per cent.

$$0.51 \times 1.2 = 0.61\Omega$$

Now we must add this value to the installation Z_e to find Z_s.

$$0.61 + 0.32 = 0.93\Omega$$

This is the calculated value of Z_s when the circuit is operating at its maximum capacity.

This value can now be compared directly to the maximum value of Z_s for a 20A BS EN 60898 type C circuit breaker. This value is 1.09Ω and it can be found in Table 41.3 of BS 7671.

To comply with the regulation, the actual value of 0.93Ω must be equal to or lower than the maximum value which is 1.09Ω. As we can see this value is acceptable.

Scenario 3

This is probably the most used and is certainly the most convenient as it requires one measurement and one minimal calculation. This method is known as the rule of thumb.

Example 3

The circuit is protected by a 32A type B BS EN 60898 circuit breaker with a maximum Z_s value of 1.37Ω. The Z_s for the circuit has been obtained by direct measurement and is 1.18Ω.

All that is required for this method is to look up the maximum Z_s permissible for the protective device, which from Table 41.3 can be seen as being 1.37Ω.

We must then multiply the maximum value by 0.8:

$$1.37 \times 0.8 = 1.09\Omega$$

We must now compare the measured value to the recalculated maximum value. The measured value must be equal to or lower than the recalculated maximum value.

As we can see the measured value is 1.18Ω and the recalculated value is 1.09Ω. As our measured value is the higher of the two it is not acceptable.

Method using tables from GN3 or the *On-site Guide*

To save us from having to carry out all of these calculations, the maximum Z_s tables provided in GN3 and the *On-site Guide* have already been corrected for temperature for us. All that is required is that we use an earth fault loop impedance tester to take a direct measurement of a circuit, then compare it to the Z_s value provided in either of these publications. Provided the measured value is equal to or lower than the value from the table, the circuit will be acceptable.

Note

There are two very simple methods of calculating the corrected value of Z_s for circuit breakers to BS EN 60898.

For the first one all you need to do is remember the number 35.25.

If you take the number 35.25 and divide it by the rating of the circuit breaker the resulting answer will be the corrected value for a B type CB.

For example, take a 20 type B CB:

$$\frac{35.25}{20} = 1.76\Omega$$

This differs by about 0.01Ω of the value provided in the *On-site Guide* and it will be fine to use for a reference.

Where the value for a type C CB is required all you need to do is halve the corrected value for a type B.

For example:

$$\frac{1.76}{2} = 0.88\Omega$$

And for a type D halve it again:

$$\frac{0.88}{2} = 0.44\Omega$$

The second method is just as simple but for this one we have to remember:

- Type B circuit breakers have to operate instantly at $\times 5$ their current rating.
- Type C circuit breakers have to operate instantly at $\times 10$ their current rating.
- Type D circuit breakers have to operate instantly at $\times 20$ their current rating

With this in mind we can use the formula:

$$Z_s \times I_a \leq U_0 \times C_{min}$$

If we transpose this formula we can get:

$$Z_s = \frac{U_0 \times C_{min}}{I_a}$$

We know that $U_0 \times C_{min} = 218.5$ volts.

If as an example we use a 20amp type B breaker we can see that the current required for automatic disconnection under fault conditions will be $20 \times 5 = 100A$.

Now the calculation is $\dfrac{218.5}{100} = 2.185\Omega$ which is rounded up to 2.19Ω.

To use this calculation we only need to remember to use 218.5 volts divided by the rating of the circuit breaker multiplied by 5 for type B, 10 for type C and 20 for type D.

Remember though, this will only work for BS EN 60898 circuit breakers.

Direct measurement

This method requires the use of an earth fault loop impedance tester. Where the measurement is to be taken from a socket outlet circuit the instrument can simply be plugged in and a resistance measurement taken. As with all live tests, leads to GS38 must be used and the lead which is supplied with the earth loop tester will be compliant.

The Z_s measurement should be taken from the point on the circuit which is the furthest from the consumer unit. It is important that the highest value is recorded and this will be at the point with the longest cable length. Of course if the installation is new and the person who is carrying out the test was also the installer, the furthest point will be probably be known. However on an existing installation the furthest point may not be known, and it may be necessary to conduct the test at several points to get the highest reading.

A circuit incorporating a socket outlet on a ring or a radial

Step 1

Use an earth fault loop impedance instrument set it onto 20Ω (unless you have a self ranging instrument).

Step 2

Ensure that all protective earthing and protective bonding is connected.

Step 3

Plug in the instrument and record the reading (Figure 5.17).

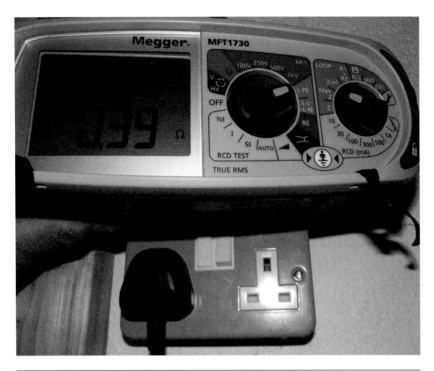

Figure 5.17 Measured value

Performing the test on a radial circuit other than a socket outlet

Earth fault impedance test on a lighting circuit (Z$_s$)
Video footage is also available on the companion website for this book.

Step 1
Ensure earthing and protective bonding is connected.

Step 2
Safely isolate circuit to be tested.

Step 3
Remove accessories at the points of the circuit at which the test is to be carried; if you are familiar with the circuit then this can be the furthest point only.

Step 4
Use an earth fault loop impedance instrument, but rather than use the lead with a plug on, you must now use the lead with three connections. This is often referred to as the fly lead.

Place the leads on correct terminals. If you are using a two lead instrument, it is as shown in Figure 5.18. If you are using a three

Figure 5.18 Two lead connection

lead instrument connect as shown in Figure 5.19 (always read the instrument manufacturer's instructions).

Step 5
Energise the circuit and record the value.

Step 6
Isolate the circuit and remove the leads.

Where the test is carried out on a lighting circuit, it can be at the ceiling rose or the switch, whichever is the most convenient.

If a two lead instrument is being used, place the probes as shown in Figure 5.20. This will also prove polarity if the switch is operated whilst carrying out the test. (This may be easier with two people.)

If a three lead instrument is being used, then connect the probes as shown in Figure 5.21 (always read the instrument instructions).

Wherever possible, always safely isolate the circuit being tested before connecting the leads.

The values measured must be compared against the maximum Z_s values which can be found in BS 7671.

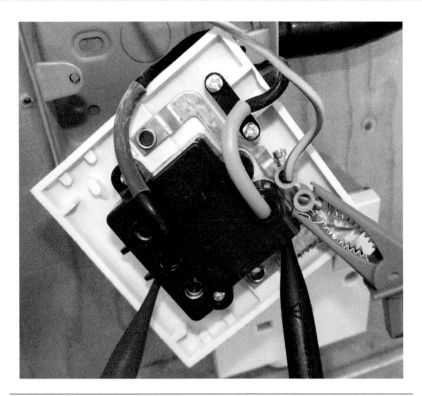

Figure 5.19 Three lead connection

Figure 5.20 Two lead connection

Figure 5.21 Three lead connection

Example 4

A ring final circuit is protected by a 30A BS 3036 semi enclosed rewirable fuse and the measured Z_s is 0.97Ω.

As this is a ring final circuit the disconnection time has to be 0.4 seconds. From Table 41.2 in BS 7671 the maximum Z_s for a 30A rewirable fuse is 1.04Ω.

The rule of thumb can now be applied which means that 80 per cent of this value must now be calculated. This can be achieved by multiplying it by 0.80.

$$1.04 \times 0.8 = 0.83Ω$$

The measured value for the circuit must now be lower than the corrected value if it is to comply with BS 7671.

Measured value 0.97Ω

Corrected value 0.83Ω

The measured value of Z_s is higher, therefore the circuit will *not* comply.

When recording Z_s for a circuit, measuring Z_e and then adding $R_1 + R_2$ is the preferred method because it will give an accurate value whereas direct measurement will include parallel paths and because of this will often give lower readings.

$Z_e + R_1 + R_2$ should always be used for an initial verification as the first recorded value will be used as a benchmark to be compared with results taken in future periodic tests.

If on a periodic inspection when using direct measurement, a higher test result is obtained than on the initial verification, it would indicate that the circuit is deteriorating and that further investigation would be required.

The methods that have been described must be fully understood by anyone who is intending to sit the City and Guilds 2394 or 5 exam on inspection and testing of electrical installations.

However! As previously described, providing that the cables used are thermoplastic or thermosetting to the required BS, the measured Z_s can be compared directly with Tables A1 to A4 in GN 3 or Tables B1 to B6 in the *On-site Guide*. These tables are already corrected for conductor operating and ambient temperature. They can also be used where the CPC is a different cross sectional area than the live conductors.

These values are pretty much the same as those calculated using the rule of thumb method as they are at approximately 80 per cent of values given in BS 7671 and are perfectly acceptable. (Remember the previous methods described *must* be understood.)

Example 5

A circuit supplying a fixed load is protected by a 20A BS 3036 fuse and the measured Z_s is 1.47Ω. The circuit C.P.C. is 1.5mm².

As this circuit is supplying a fixed load and does not exceed 32A the maximum permitted disconnection time is 0.4 seconds.

The table that should be used is B1 (i) from the *On-site Guide*. Using the table it can be seen that for a 20A device protecting a circuit with a 1.5mm² CPC the maximum permissible Z_s is 1.3Ω. As the measured Z_s is higher than the maximum permitted the circuit will not comply with the requirements of the regulations.

Example 6

A circuit supplying a cooker outlet protected by a 45A BS 1361 fuse has a measured Z_s of 0.4Ω. The CPC is 4mm².

As this circuit is rated at 45A the maximum disconnection time would be 5 seconds.

The table to use for this circuit is B5 (ii) and the maximum permitted Z_s is 0.67Ω. This circuit will comply.

Example 7

A circuit supplying a lighting circuit is protected by a 6A type C BS EN 60898 protective device which has a measured Z_s of 2.9Ω.

It should be remembered that miniature circuit breakers will operate at 0.1 seconds providing that the measured Z_s is equal to or lower that the values given in the tables. We do not have to worry about 0.4 or 5 second disconnection times for these devices.

The table to use for this example is B6 in the *On-site Guide*; the maximum permitted Z_s is 2.93Ω. The measured value of 2.9Ω is lower than the maximum permitted therefore this circuit would comply.

Remember the values in GN3 and the *On-site Guide* are corrected for temperatures of 10°C and no other calculation is required providing the circuit has not been installed using Table 7.1 of the *On-site Guide*.

If the ambient temperature is below or above 10°C then correction factors from Table 2E of the *On-site Guide* must be used as follows.

Using the previous example 1:

A circuit supplying a fixed load is protected by a 20A. BS 3036 fuse and the measured Z_s is 0.97Ω. The circuit CPC is 1.5mm² and the *ambient temperature is 23°C*.

As this circuit is rated at less than 32A the maximum permitted disconnection time is 0.4 seconds.

The table that should be used is B1(i) from the *On-site Guide*. Using the table it can be seen that for a 20A device protecting a circuit with a 1.5mm² CPC the maximum permitted Z_s is 1.3Ω.

Now the temperature has to be taken into account.

Using Table B8 from Appendix B of the *On-site Guide* it can be seen that the nearest value to the temperature which was measured (23°) is 25°C. (Always round up to be on the safe side.) The correction factor for 25°C is 1.06.

This value (1.06) is now used as a multiplier to the maximum permitted Z_s (1.3Ω) to calculate the maximum Z_s for the circuit at 25°C.

$$1.3 \times 1.06 = 1.38Ω$$

This is the maximum measured Z_s permissible for the circuit at 25°C.

When carrying out earth fault loop testing on circuits which are not protected by residual current devices, the process is very simple and it is very unlikely that the test will trip any circuit breakers other than possibly a type B 6A BS EN 60898. This device may trip because the test current used for loop testing is normally 25A and type B circuit breakers must operate within a window of between 3 and 5 times its operating current. This of course means that the 6A device will operate at between 18 and 30 amps and the test current may trip them.

Most earth fault loop testers have a low current no trip setting which allows the test to be completed without tripping an RCD. On occasion though, particularly on older installations which have an RCD as a main switch, even a low current test will trip the RCD due to very small earth leakage currents being in the system which do not amount to the trip rating of the RCD. When the RCD test is carried out, the test current added to the current already in the system can be enough to trip the RCD.

In installations which are protected by a single RCD and you need to be absolutely sure that the RCD does not trip, or where you do not have a low trip current tester, the following method can be used to measure Z_s.

Earth loop impedance using a high current loop test instrument without tripping an RCD

It can be very inconvenient if an RCD is tripped by accident. Most electricians will have tripped an RCD which is being used to protect the whole installation while using a D lock or low current instrument; this can be very embarrassing and very inconvenient, particularly if it requires the resetting of time clocks and other electronic equipment.

Other electricians will not have either of these instruments and have to rely on the calculation $Z_s = Z_e + R_1 + R_2$. This is fine and perfectly acceptable.

Sometimes it is more satisfying to carry out a live test that will give a direct reading of Z_s to ensure that no loose connections or high resistance joints are affecting the circuit. Some electricians just prefer the simplicity of a live test as it leaves very little to chance.

This is a very simple process and it can be carried out as follows:

- Isolate the circuit to be tested.
- Link phase and earth at the furthest point of the circuit using a lead with a crocodile clip on each end (Figure 5.22). If it is a socket outlet then a plug top with earth and phase linked can be used (it is advisable to clearly mark the plug top).
- Using a high current earth fault loop impedance test instrument, place one probe (*black*) on to the isolated terminal of the circuit protective device (Figure 5.23).
- Place the other probe (*red*) on to the incoming phase of the RCD or main switch (Figure 5.23).
- Operate instrument and record the result.
- This will be Z_s for the circuit and the RCD will not have tripped.

If your test instrument is a three lead instrument connect the black and green leads together onto the isolated terminal.

Figure 5.22 Line and earth linked

Figure 5.23 Incoming supply to outgoing line

Prospective fault current test (I$_{pf}$)

This is a live test and great care should be taken.

A prospective fault current tester is normally combined with an earth loop impedance tester. The measured value is normally shown in kA (kilo amps).

Regulations 434.1 and 643.7.3.201 require that the prospective fault current is determined at every relevant point of the installation. This may be at the origin of the supply or at the end of every distribution circuit. Prospective fault current (I$_{pf}$) is the highest current which could flow in an installation at the point at which it is measured.

Depending on the type of installation the highest value could be either between live conductors or live conductors and earth.

To obtain the value of prospective fault current (I$_{pf}$), we must first determine the value of the prospective short circuit current.

Prospective short circuit current (PSCC) is the maximum current that could flow between phase and neutral on a single phase supply or between phase conductors on a three phase supply.

Prospective earth fault current (PEFC) is the maximum current that could flow between live conductors and earth.

Prospective fault current test

Video footage is also available on the companion website for this book.

The higher of these values is known as prospective fault current.

The highest prospective fault current will be at the origin of the installation and must be measured as close to the meter position as possible, usually at the main switch for the installation. It is measured between phase and neutral.

This can be done by:

- enquiry to the supplier
- calculation
- measurement.

Enquiry

This is a matter of a phone call to the electricity supplier of the installation. They will tell you the maximum PFC. Usually this is a lot higher than the value will actually be, but if you use this value you will be on the safe side.

Calculation

The I_{pf} can only be calculated on a TNCS system. This is because the neutral of the supply is used as a protective earth and neutral conductor (PEN).

When the earth fault loop impedance is measured the value measured is in ohms. To convert this value to prospective short circuit current we must use the following equation:

$$\text{PSCC} = \frac{V}{Z_e} = I$$

It is important to remember that the line voltage to earth of the supply transformer is used U_0 230 v (BS 7671 Appendix 2).

Example 8

Z_e is measured at 0.28Ω.

$$\frac{230}{0.23} = 1000A$$

A useful tip is that when you have measured Z_e on a TNCS system, set your instrument to PFC and repeat the test. This will give you the value for PFC (I_{pf}) and save you doing the calculation.

Measurement

This is carried out using a prospective short circuit current tester. As with all tests it is important that you have read the instructions for the instrument which you are going to use.

If you are using a two lead instrument with leads and probes to GS38:

Step 1

Set instrument to PFC (Figure 5.24)

Step 2

Place the probes on the phase and neutral terminals at supply side of the main switch (Figure 5.25)

Step 3

Operate the test button and record the reading (Figure 5.26).

When carrying out the test using a three lead instrument with leads to GS38, it is important that the instrument instructions are read and fully understood before carrying out this test.

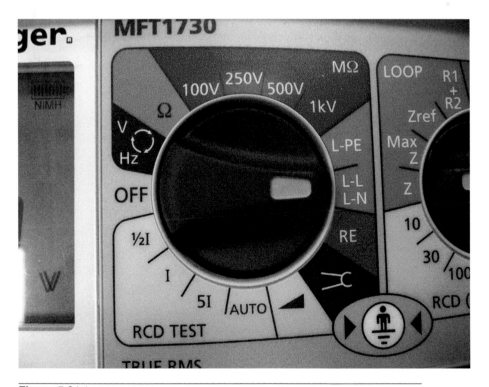

Figure 5.24 Instrument set

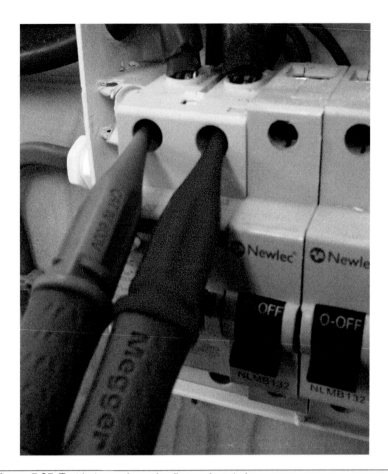

Figure 5.25 Test between incoming line and neutral

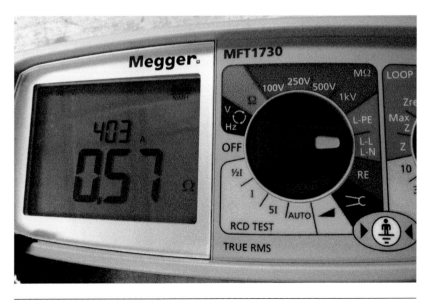

Figure 5.26 Measured value

Three lead test

Step 1

Place the line and N lead on the supply side of the main switch and the neutral and earthing probes/clips onto the earthing terminal (Figure 5.27).

Step 2

Operate the test button and record the reading.

If the supply system is a three phase and neutral system then the highest current that could flow in it will be between lines. Some instruments will not be able to measure the high current that would flow under these circumstances.

Under these circumstances the measurement should be made between any phase and neutral at the main switch and the measured value should be *doubled*.

For your personal safety and the protection of your test equipment it is important to read and fully understand the instructions of your test instrument before commencing this test.

Some PSCC instruments give the measured value in ohms, not kA. If this is the case a simple calculation, using Ohm's law, is all that is required.

Figure 5.27 Leads connected

Figure 5.28 BS EN 60898

Example 9

Measured value is 0.07Ω.

Remember to use U_0 in this calculation (230v).

$$PSCC = \frac{230}{0.07} = 3285A$$

It is important that the short circuit capacity of any protective devices fitted exceeds the maximum current that could flow at the point at which they are fitted.

When a measurement of I_{pf} is taken as close to the supply intake as possible, and all protective devices fitted in the installation have a short circuit capacity that is higher than the measured value, then Regulations 432.1 and 432.3 will be satisfied.

In a large installation where distribution circuits (sub mains) are used to supply distribution boards it can be cost effective to measure the I_{pf} at each board. The I_{pf} will be smaller and could allow the use of a protective device with a lower short circuit rating, as these will usually be less expensive.

Table 7.2.7 (i) in the *On-site Guide* gives rated short circuit capacities for devices; examples of these values are provided in Table 5.3 and can also be obtained from manufacturers' literature.

Remember that a BS 1361 fuse is referred to in BS 7671 as a BS 88 – 3 type C.

Circuit breakers to BS 3871 are marked with values M1 to M9 – the number indicates the maximum value of kA that they are rated at.

Circuit breakers to BS EN 60898 and RCBOs to BS EN 61009 show two values in boxes usually on the front of the device (Figure 5.28).

The square box will indicate the maximum current that the device could interrupt and still be reset.

Ics rating   3̲

Table 5.3 Examples of rated short circuit capacities

Examples	Rated short circuit capacity
Semi enclosed BS 3036	1kA to 4kA depending on type
BS 1361 Type 1 Type 2	16.5kA 33kA
BS 88–2.1	50kA at 415 volts
BS 88–6	16.5kA at 240 volts 80kA at 415 volts

The rectangular box will indicate the maximum current that the device can interrupt safely.

Icn rating | 6000 |

If a value of fault current above the rated Isc rating of the device were to flow in the circuit, the device would no longer be serviceable and would have to be replaced.

A value of fault current above the Icn rating would be very dangerous and possibly result in an explosion causing major damage to the distribution board/consumer unit.

Functional testing

All equipment must be tested to ensure that it operates correctly. All switches, isolators and circuit breakers must be manually operated to ensure that they function correctly, and also that they have been correctly installed and adjusted where adjustment is required.

Video footage is also available on the companion website for this book.

Residual current device

The instrument used for this test is an RCD tester, and it measures the time it takes for the RCD to interrupt the supply of current flowing through it. The value of measurement is either seconds or milliseconds.

Before we get on to testing let's consider what types of RCDs there are, what they are used for and where they should be used.

Types of RCD

Voltage operated

Voltage operated ELCBs (earth leakage current breakers) are not uncommon in older installations. This type of device became obsolete in the early 1980s and must *not* be installed in a new installation or alteration as they are no longer recognised by BS 7671.

They are easily recognised (Figure 5.29) as they have two earth connections, one for the earth electrode and the other for the installation earthing conductor. The major problem with voltage operated devices is that a parallel path in the system will probably stop it from operating.

These types of devices would normally have been used as earth fault protection in a TT system.

Figure 5.29 Voltage operated RCD

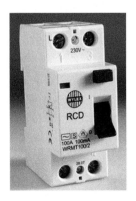

Although the electrical wiring regulations BS 7671 can not insist that all of these devices are changed, if you have to carry out work on a system which has one it must be replaced to enable certification to be carried out correctly. If however a voltage operated device is found while preparing a periodic inspection report a recommendation that it should be replaced would be the correct way of dealing with it.

BS 4293 General purpose device

These RCDs (Figure 5.30) are very common in installations although they ceased to be used in the early 1990s. They have been replaced by BS EN 61008-1, BS EN 61008-2-1 and BS EN 61008-2-2.

They are used as stand-alone devices or main switches fitted in consumer units/distribution boards

This type of device provides protection against earth fault current. They will commonly be found on TT systems 15 or more years old, although they may be found on TNS systems where greater protection was required.

The problem with using a low tripping current device as the main switch is that nuisance tripping could occur. This type of protection would not be acceptable as compliance with the 17 edition of BS 7671 Wiring Regulations. If major alterations were being carried out then the protection would need to be changed to comply with the modern way of thinking which is explained later in this chapter.

BS 4293 type S

These are time delayed RCDs (Figure 5.31) and are used to give good discrimination with other RCDs.

BS EN 61008-1 General purpose device

This is the current standard for a residual current circuit breaker (RCCB) and provides protection against earth fault current (Figure 5.32). These devices are generally used as main switches in consumer units/distribution boards.

Three phase devices are also very common (Figure 5.33).

BS 7288

This is the current standard for RCD protected socket outlets and provides protection against earth fault current (Figure 5.34). Where the socket outlets are sited outside, waterproof BS 7288 outlets are used to IP 56.

BS EN 61009-1

This is the standard for a residual current circuit breaker with overload protection (RCCBO) (Figure 5.35).

These devices are generally used to provide single circuits with earth fault protection, overload protection and short circuit protection. They are fitted in place of miniature circuit breakers and the correct type should be used (type B, C or D).

BS EN 61008-1 type S

These are time delayed RCDs and are used to give good discrimination with other RCDs (Figure 5.36).

Section 3 of the *On-site Guide* gives good examples of how these devices should be used within an installation.

RCDs and supply systems

TT system

BS 7671 states that care must be taken to ensure that the operation of a single protective device should not cause a dangerous situation (Regulation 314.2). One RCD protecting the whole installation is now no longer acceptable in the majority of installations.

Compliance with Regulation 314 can be achieved by using a split board with a non-RCD main switch and RCDs protecting both sides of the split board. In this instance careful consideration should be given to how the circuits are divided, possibly mixing upstairs and downstairs circuits to each side of the board. This would avoid the whole of the upstairs or downstairs circuits having loss of supply due to a fault on a single circuit.

Another method would be to use a consumer unit with a main switch to BS EN 60947-3 and RCBOs to BS EN 610091 as protective devices

Figure 5.32 Single phase

Figure 5.33 Three phase

Figure 5.34 RCD socket outlet

for all circuits. This option is perfectly satisfactory but can work out a little expensive!

TNS and TNCS systems

The previous options would be suitable for these systems but where RCD protection is not required for all circuits a standard board with a non-RCD protected main switch could be used. RCCBOs would then be fitted for the circuits requiring RCD protection and normal protective devices used for any circuits not requiring RCD protection.

I am sure that many other options are or will become available due to different products being introduced by the many manufacturers of electrical equipment

Testing of RCDs

Remember that these are live tests and care should be taken whilst carrying them out.

The instrument to be used to carry out this test is an RCD tester, with leads to comply with GS38.

Voltage operated (ELCBs)

No test is required as they should now be replaced.

BS 4293 RCDs

If this type of RCD is found on TT systems or other systems where there is a high value of earth fault impedance (Z_e), the RCD tester should be plugged into the nearest socket or connected as close as possible to the RCD. The tester should then be set at the rated tripping current of the RCD ($I_{\Delta n}$), in this example, it's rated at 30mA. (Be careful and do not mistake the tripping current for the current rating of the device.)

Step 1

The test instrument must then be set at 50 per cent of the tripping current (15mA) (Figure 5.37).

Step 2

Push the test button of the instrument – the RCD should not trip (Figure 5.38).

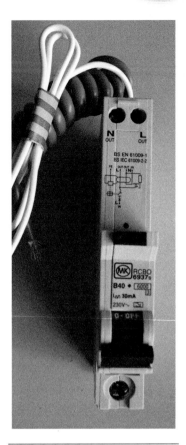

Figure 5.35 BS EN 61009-1

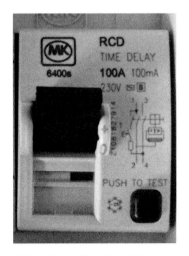

Figure 5.36 BS EN 61008-1 type S

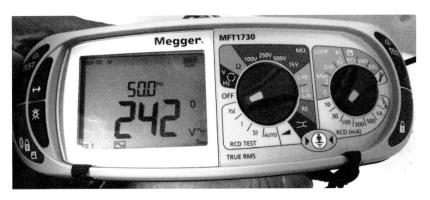

Figure 5.37 Set at times half

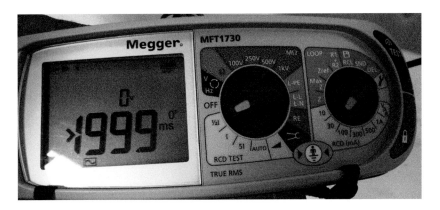

Figure 5.38 No trip

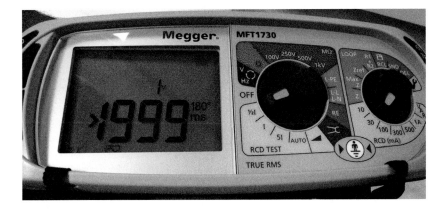

Figure 5.39 No trip

Step 3

The test instrument will have a switch on it which will enable the instrument to test the other side of the waveform 0° ∿ 180°. This switch must be moved to the opposite side and the test repeated (Figure 5.39) Again the RCD should not trip.

If during testing of any RCD it trips during the 50 per cent test do not automatically assume that the RCD is at fault.

Consider the possibility that there is a small earth leakage on the circuit or system. Switch all circuits off and test the RCD on the load side at 50 per cent using fly leads. If it still trips, the RCD should be replaced.

If it does not trip, turn each circuit on one at a time, carrying out a 50 per cent test each time a circuit has been turned on. When the RCD trips, switch off all circuits except the last one which was switched on. Test again. If the RCD trips, carry out an insulation test on this circuit as it probably has a low insulation resistance. If the RCD does not trip it could be an accumulation of earth leakage from several circuits and they should all be tested for insulation resistance.

Step 4

Now set the test current to the rated tripping current (30mA) (Figure 5.40).

Push the test button, the RCD should trip within 200 milliseconds.

Step 5

Reset the RCD.

Step 6

Move the waveform switch to the opposite side, and repeat the test, again it must trip within 200 milliseconds (Figure 5.41).

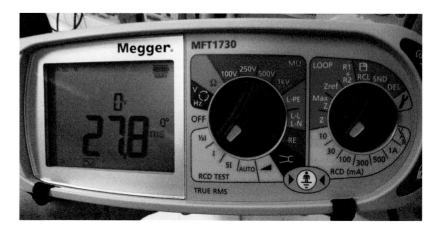

Figure 5.40 Test at times one

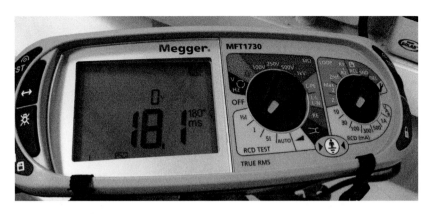

Figure 5.41 Test at 180°

Step 7

Reset the RCD and the slowest time in which it tripped should be entered on to the test result schedule.

Step 8

Set the test current to 5 times the rated tripping current (150mA) (Figure 5.42).

Step 9

Push the test button and the RCD should trip within 40 milliseconds (Figure 5.43).

Step 10

Move the waveform switch to the opposite side, and repeat the test (Figure 5.44). Again it must trip within 40 milliseconds (5 times faster than the times 1 test).

After completion of the instrument tests

Step 11

Push the integral test button on the RCD to verify that the mechanical parts are working correctly (Figure 5.45).

Step 12

Ensure that a label is in place to inform the user of the necessity to use the test button quarterly (Figure 5.46).

The 5 times test must only be carried out on RCDs with trip ratings ($I_{\Delta n}$) up to 30mA.

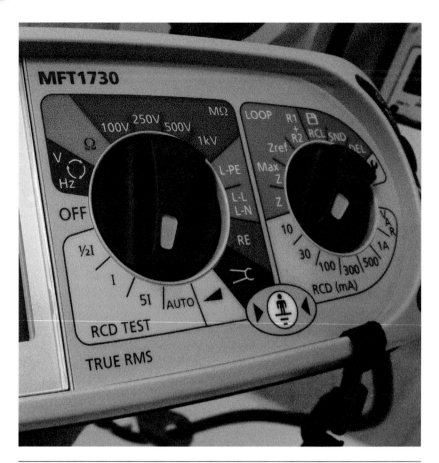

Figure 5.42 Set at times five

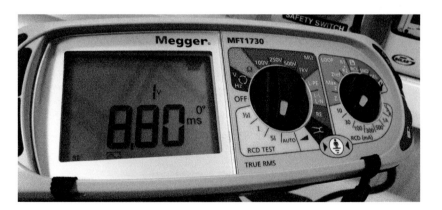

Figure 5.43 Test at zero degrees

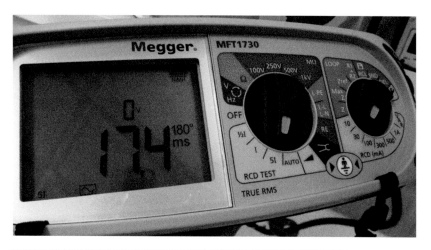

Figure 5.44 Test at 180°

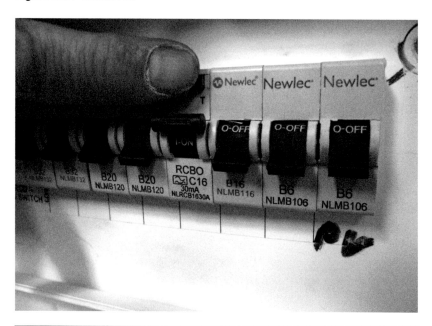

Figure 5.45 Manual test

This installation, or part of it, is protected by a device which automatically switches off the power supply if an earth fault develops. **Test quarterly** by pressing the button marked **'T'** or **'Test'**. The device should switch off the supply and then should be switched on to restore the supply. If the device does not switch off the supply when the button is pressed seek expert advice.

Figure 5.46 Test label

BS EN 610081

These devices should be tested in exactly the same manner as BS 4293 using the same test instrument. However, the difference is that the tripping time when carrying out the 100 per cent test is increased to 300 milliseconds.

BS 4293 type S

This device has a built in time delay. The simple way to think about them is that they do not recognise a fault for 200 milliseconds, and they must trip within 200 milliseconds after that.

Step 1

Plug in or connect the RCD as close as possible to the RCD to be tested.

Step 2

Set the instrument on the trip current of the RCD and ensure that it is set for S type.

Step 3

Test at 50 per cent and the device should not trip.

Step 4

Repeat the test on opposite waveform.

Step 5

Set the test instrument on 100 per cent and carry out the test. The RCD should trip within 400 milliseconds (200 ms time delay and 200 ms fault).

Step 6

Repeat on opposite wave form.

The slowest operating time at 100 per cent test should be recorded, as should the fact that it is a S type.

BS EN 61008 type S

This device has a time delay of 200 milliseconds and a tripping time of 300 milliseconds, making a maximum tripping time of 500 milliseconds.

The test should be carried out as the BS 4293 type S but remember the different tripping time.

BS 7288 RCD protected socket

This device should be tested the same as a BS 4293 and the tripping times are the same.

Consideration should be given to whether the socket will supply portable equipment outdoors. If it can it should be tested at 5 times its rating.

BS EN 61009 RCBOs

These devices should be tested as BS 4239 RCDs but the disconnection times are:

- 50 per cent test on both sides of wave form, no trip.
- 100 per cent test on both sides of wave form, must trip within 300 milliseconds.
- If used as supplementary protection, the 5 times test must also be carried out, it must trip within 40 milliseconds.

The product standard performance criteria can be found in Table 3A which is in Appendix 3 of BS 7671.

Always ensure that it is safe to carry out these tests.

Remember to remove any loads and ensure that the disconnecting of the supply due to the test will not affect any equipment or cause damage.

If any people are within the building ensure that they are aware of testing being carried out, and that a loss of supply is likely.

CHAPTER 6

Completion of test certificates

Minor electrical installation works certificate

A minor works certificate (Figure 6.1) is to be completed when additions or alterations to existing circuits have been carried out, for example, an additional lighting point or socket outlet.

If the alteration results in the protective device being changed or an RCD installed, then an electrical installation certificate is required, not a minor works certificate.

If more than one circuit has been added to, then a separate minor works certificate must be issued for each modification.

These certificates vary slightly depending on which certification body has supplied them; some require slightly more information than others.

For any candidates who are studying for any of the City and Guilds courses such as the 2394 or 5, it is worthwhile remembering that the course is based on the documents from BS 7671.

The minor electrical installation works certificate consists of 4 parts.

Part 1 is the description of the minor works and requires the following five pieces of information:

1 Details of the client and date on which the work was completed.
2 The installation address or the location at which the work has been carried out.
3 Description of the minor work carried out. It is important to document exactly what you have done.
4 Details of departures. This would usually be anything which does not meet British standards such as a new invention, or something which has been made for a special purpose. A wrought iron lamp which has been made by a blacksmith would be a good example.

MINOR ELECTRICAL INSTALLATION WORKS CERTIFICATE

To be used only for minor electrical works which do not include the provision of a new circuit

---

Part 1 : Description of the minor work carried out

1. Details of the client.. Date work completed...............................

2. Installation address or location..

3. Description of the minor work carried out...

4. Details of any departures from BS 7671: 2018 for the circuit worked on. See regulations 120.3, 133.1.3 and 133.5. Where applicable a suitable risk assessment must be completed and attached to this document.

   Risk assessment attached ☐

5. Comments on the existing installation (including any defects observed. See regulation 644.1.2)

   ....

Part 2: Presence and suitability of the installation earthing and bonding arrangements (see regulation 132.16)

1. System earthing arrangement : TT ☐   TN-S ☐   TN-C-S ☐

2. Earth fault loop impedance (Z_{db}) at the distribution board supplying the final circuit.................Ω

3. Presence of adequate main protective conductors:

Earthing conductor ☐

Main protective bonding conductor(s) to: Water ☐   Gas ☐   Oil ☐   Structural steel ☐ Other......................... ☐

Part 3: Circuit details

DB reference No:............. DB location and type ....

Circuit No:........................ Circuit description...

Conductor sizes: Live conductors...............mm^2 CPC...................mm^2

Circuit overcurrent protective device: BS(EN).................   Type................Rating...................A

Part 4: Test results for the altered or extended circuit (where relevant and practicable)

Protective conductor continuity: R_2 .... Ω   or $R_1 + R_2$...........................Ω

Continuity of ring final circuit conductors: L/L............Ω   N/N..............Ω   CPC/CPC..............Ω

Insulation resistance: Live - Live...............MΩ   Live - Earth..............MΩ

Polarity correct: ☐   Maximum measured earth fault loop impedance: Z_s ....Ω

RCD operation: ☐   Rated residule operating current ($I_{\Delta n}$)......................mA

Disconnection time ....ms   Test button operation is satisfactory ☐

Part 5: Declaration

I certify that the work covered by this certificate does not impair the safety of the existing installation and that the work has been designed, constructed, inspected and tested to the best of my knowledge and belief, at the time of my inspection complied with BS 7671 except as detailed in Part 1 of this document.

Name..   Signature...

For and on behalf of...   Position...

Address...   Date..

Figure 6.1 Minor works certificate

5 Comments on the existing installation. Any defects found which do not have an influence on the work which you have carried out but which you have identified as being dangerous should be noted here. Any defects which affect the circuit which you are adding to should be rectified before completing the work.

Part 2 Presence and suitability of the installation earthing and bonding arrangements.

1 System earthing arrangement. Is it TT, TN-S or TN-C-S?

2 What is the measured Zs at the distribution board which the circuit you are adding to/altering is supplied from?

3 Has the installation got a suitable main protective earthing and bonding conductor? (This does not require the upgrading of earthing and bonding as long as the circuit which you have altered is safe).

Part 3 Circuit details

- Number of the DB and its location.
- Circuit identification: number and what it does.
- Conductor sizes for live conductors and CPC.
- Type of overcurrent protective device and its rating.

Part 4 Test results for the completed circuit which has been altered or extended.

- Common sense has to prevail here, not everything listed has to be completed. It may be that R2 or even R1 + R2 cannot be measured, but presence of a suitable earth will be confirmed by the loop test.
- Continuity of CPC, this can be by method 1 or 2 where possible.
- Ring final circuit conductors measured end to end. Always required to prove ring continuity.
- Insulation resistance, generally between live conductors joined together and earth as the circuit may have loads connected.
- Measured earth fault loop at the point which you have altered the circuit and also at the furthest point, the highest value should be recorded.
- Polarity check, and RCD test.

Electrical installation certificate

This certification that is required for a new installation or circuit.

The electrical installation certificate (Figure 6.2) is to be completed for a new circuit, a new installation, a rewire and any circuit where the protective device has been changed.

ELECTRICAL INSTALLATION CERTIFICATE
(REQUIREMENTS FOR ELECTRICAL INSTALLATIONS – BS 7671 IET wiring regulations)

Details of the client

Installation address

Description and extent of the installation Description of the installation	New Installation	☐
Extent of the installation covered by this certificate:	Addition to an existing installation	☐
	Alteration to an Existing installation	☐

For Design

I/We being the person(s) responsible for the design of the electrical installation (as shown by the signatures below). The particulars of which are described above, having exercised reasonable care and skill when carrying out the design and also where the certificate applies to an addition or alteration, the safety of the existing installation is not compromised. I/We hereby certify that the design work for which I/we have been responsible is to the best of my belief and knowledge in accordance with BS 7671: 2018, amended to .... (date) except for any departures as detailed below.

Details of departures from BS 7671 (Regulations 120.3, 133.1.3 and 133.5)

Details of permitted exceptions (reg 411.3.3). Where applicable a suitable risk assessment must be completed and attached to this certificate

Risk assessment attached ☐

The extent of the liability is limited to the work described as the subject of this certificate

For the design of the installation:

Signature .... Date .... Name (in Block Letters)... Designer No 1

Signature .... Date .... Name (in Block Letters)... Designer No 2 where applicable

For Construction

I being the person(s) responsible for the construction of the electrical installation (as shown by the signatures below). The particulars of which are described above, having exercised reasonable care and skill when carrying out the construction. I hereby certify that the design work for which I have been responsible is to the best of my belief and knowledge in accordance with BS 7671: 2018, amended to .... (date) except for any departures as detailed below.

Details of departures from BS 7671 (Regulations 120.3 and 133.5)

The extent of the liability is limited to the work described as the subject of this certificate.

For **Construction** of the installation: Signature.. Date ....

Name (In Block Letters)..For construction

For Inspection and Testing

I being the person(s) responsible for the inspecting and testing of the electrical installation (as shown by the signatures below). The particulars of which are described above, having exercised reasonable care and skill when carrying out the Inspecting and Testing. I hereby certify that the design work for which I have been responsible is to the best of my belief and knowledge in accordance with BS 7671: 2018, amended to .... (date) except for any departures as detailed below.

Details of departures from BS 7671 (Regulations 120.3 and 133.5)

The extent of the liability is limited to the work described as the subject of this certificate.

For **Inspection and Testing** of the installation: Signature.. Date ....

Name (In Block Letters)..Inspector

Next Inspection

I/We being the designer(s) recommend that this installation is subjected to a periodic inspection and test after an interval of not more than ....years/months

---

Figure 6.2a Electrical installation certificate

Particulars of Signatories to the electrical installation certificate

Designer

Name: .... Company:..

Address: ....

.... Postcode: .... Telephone No: ....

Constructor

Name: .... Company:..

Address: ....

.... Postcode: .... Telephone No: ....

Inspector

Name: .... Company: ....

Address: ....

.... Postcode: .... Telephone No: ....

Supply Characteristics and Earthing Arrangements

Earthing arrangements	Number and type of live conductors		Nature of Supply Parameters	Supply Protective Device
TT ☐ TN-S ☐ TN-C-S ☐ TN-C ☐ IT ☐	AC ☐ 1-phase, 2 wire ☐ 2-phase, 3 wire ☐ 3-phase, 3 wire ☐ 3-phase, 4 wire ☐	DC ☐ 2-wire ☐ 3-wire ☐ other ☐	Nominal Voltage, U /U$_o$ [1] ....V Nominal frequency, f [1] ....Hz Prospective fault current, I$_{pf}$ [2] ....kA External loop impedance, Z$_s$ [2]................Ω Note (1) by enquiry (2) By measurement or enquiry	BS (EN) .... Type.. Rated Current..A

Other sources of supply (As attached information)

Particulars of The Installation Referred to on the Certificate

Means of Earthing	Maximum Demand
Distributors facility ☐ Installation earth electrode ☐	Maximum Demand load..kVA/Amps (delete as necessary) **Details of the installation earth electrode if installed** Type (e.g. Rods, tape, plate etc)... Location... Electrode resistance to earth.....................Ω

Main Protective conductors

Earthing conductor	Material................................ csa ....mm^2	Connection / continuity verified ☐
Main protective bonding conductors (to extraneous conductive parts)	Material............................csa........................mm^2	Connection / continuity verified ☐

To water installation pipes ☐   To gas installation pipes ☐   To oil installation pipes ☐   To structural steel ☐   To lighting protection ☐

To other ☐ ....

Main Switch / Fuse / Circuit Breaker / RCD

Location... .... BS(EN).. Number of poles.....................................	Current rating...A Fuse / device rating................................A Voltage rating V	Where RCD is main switch Rated residual operating current (I$_n$)....................mA Rated time delay..ms Measured operating time ....ms

Comments On The Existing Installation (Where this is an addition or alteration see regulation 644.1.20)

....

....

....

....

....

....

Schedules

The attached schedules form part of this document and this circuit is only valid when they are attached to it

Number of schedules of inspection...Number of schedules of test results ....are attached

Figure 6.2b Continued.

In the case of a consumer unit change only, an electrical installation certificate would be required for the consumer unit and a periodic inspection report should be completed for the existing installation.

A standard electrical installation certificate can be used for any installation. However if the work to be certificated is covered by Building Regulations Part P, certificates are available solely for this purpose. These certificates simplify the paperwork by including a schedule of inspection and a schedule of test results on the same document and are usually purchased from the registration body with whom you register.

A schedule of test results (Figure 6.3) and a schedule of inspection must be completed to accompany an electrical installation certificate and a periodic inspection report. A schedule of inspection can be found in BS 7671.

These certificates vary slightly depending on which certification body has supplied them; some require slightly more information than others. Figures 6.2 and 6.3 are typical of an electrical installation certificate.

The information required is as follows.

- *Details of client*. Name and address of the person ordering the work.
- *Location/address*. The address at which the work is carried out. The name of the occupier.
- *Description and extent of the installation*. What part of the installation does this certificate cover, is it all of the installation, or is it a single circuit. It is vital that this part of the certificate is completed as accurately as possible.

There are generally three tick boxes regarding the nature of the installation.

- *New installation*. To be ticked if the whole installation is new, this would include a rewire.
- *Addition to an existing installation*. This box would be ticked when the work is added to an existing installation, this could be a single new circuit or perhaps the installation of circuits in an extension.
- *Alteration to an existing installation*. This box is used to indicate that the characteristics of an existing circuit have been altered. This would include extending/altering a circuit and changing the protective device. The replacement of consumer units and/or the fitting of RCDs would also be included under this heading.

Design, construction, inspection and testing

The person or persons responsible for each of these must sign. It could be one person or possibly three, depending on the job. However, it is important that all boxes have a signature.

GENERIC SCHEDULE OF TEST RESULTS

DB Reference no........................
Location................................
Z_s at DB (Ω).......................
I_{pf} at DB (kA)......................
Correct supply polarity confirmed ☐
Phase sequence confirmed (where appropriate)

Details of circuits and/or installed equipment which may be vulnerable to damage when testing...

Details of test instruments used (state serial or asset numbers)
Continuity.............................
Insulation resistance.................
Earth fault loop impedance............
RCD...................................
Earth electrode resistance............

Tested by:
Name in capitals......................
Signature....................Date..........

Test Results

Circuit number	Circuit description	BS (EN)	Type	Rating (A)	Breaking capacity (kA)	RCD I$_{\Delta n}$ mA	Maximum permitted Z_s	Ring final circuit continuity (Ω) r1 (Line)	rn (neutral)	r2 (cpc)	Continuity (Ω) ($R_1 + R_2$) or R_2	R2	Insulation resistance test Volts	Insulation resistance (MΩ) Live - Live	Live - Earth	Polarity	Z_s (Ω) Maximum measured	RCD Disconnection Time (ms)	RCD test button operation	AFDD Manual AFDD test button operation	Remarks Use separate sheet if required

Where the tabulated values of maximum permitted earth fault loop impedance is taken from a source other than chapter 41 of BS 7671, state the source of the information in the remarks column of this schedule

Figure 6.3 Schedule of test results

Usually in this area there will be two boxes referring to BS 7671. To complete this correctly look at the top right hand corner of BS 7671 – BS 7671 'Year' can be seen and just below it the date of the amendments; this will indicate the most recent amendment.

This section also requires that any departures are recorded.

Next inspection

The person who has designed the installation or the part of it that this certificate covers must recommend when the first periodic inspection and test is carried out. This will be based on the type of use to which it will be put, and also the type of environment.

Supply characteristics and earthing arrangements

Earthing
Is it TT, TNS or TNCS?

Number and type of live conductors
Usually 1 phase 2 wire or 3 phase 4 wire (only live conductors).

Nature of supply parameters
This can be gained by enquiry or measurement.

U is line to line or line to neutral

U_o *is line to earth*

Do not record measured values. If three phase or three phase and neutral then the value will be U. 400 v. If single phase then the U_o will be 240 v.

Frequency
This will normally be 50Hz although on some special installations this may change.

Prospective fault current (I_{pf})
This is the maximum current which could flow in the circuit.

External earth loop impedance (Z_e)
This should be measured between the phase and earth on the live side of the main switch with the earthing conductor disconnected. (Remember to isolate installation first.) If measurement is not possible then it can be obtained by enquiry to the supply provider.

Supply protective device

Type

Usually this will be a BS 88 or BS 1361 cartridge fuse and it will normally be marked on the supply cut out. If it is not then it should be found by enquiry to the supply provider.

Rated current

This is the current which the protective device is rated at, it will normally be found printed on the fuse carrier.

Particulars of the installation referred to in the certificate

Means of earthing

Is the earthing supplied by the distributor or has it got an earth electrode?

Maximum demand

What is the load per phase? This value is not the rating of the supply fuse or the addition of the circuit protective device ratings. It must be assessed using diversity. The use of Appendix 1 in the *On-site Guide* can be helpful, however the use of common sense and experience is probably the best way to deal with this.

This can be give in Amps or Kva (Kva = volts × amps)

Details of earth electrode

If the system has an earth electrode, what type is it? Where is it? What is its resistance? (This is usually measured with an earth loop impedance tester, using the same method as for Z_s.)

Main protective conductors

Earthing conductor

What is it made of and what size is it, has the connection been verified?

Main protective bonding conductors

What is it made of and what size is it, has the connection been verified?

Is the bonding connected to the incoming water and gas service?

Other elements

Are there any other parts of the installation which have protective bonding connected? This could be incoming oil lines, central heating etc.

Main switch or circuit breaker

* *What type of switch is it*? BS, BS EN or IEC
* *Number of poles*: single pole, double pole, triple pole or triple pole and neutral.
* *Voltage and current rating*: whatever is marked on it.
* *Location*: where is it?
* *Fuse rating or setting*: where the main switch is a switched fuse or circuit breaker.
* *Rated residual operating current $I_{\Delta n}$*: if an RCD is fitted as a *main switch*, the operating current $I_{\Delta n}$ and the operating time at $I_{\Delta n}$ must be recorded.

Comments on the existing installation

If the certificate covers the whole installation then usually 'none' will be entered here. If the installation is one that you are adding to or you have any concerns (perhaps the socket outlets or cables which are concealed in plaster are not compliant with the latest edition of BS 7671), you may enter here that a periodic inspection report would be advisable, or possibly be more specific if necessary.

Wiring regulations are not retrospective and it is not a requirement for wiring that complied when it was installed is updated. You may be the first person to have looked at the installation for many years, your professional advice could be important to your client; it is also an excellent sales opportunity.

It is important to remember that if you are completing an electrical installation certificate, then the earthing and bonding arrangements must be improved to comply with the requirements set out in the latest edition of BS 7671.

Schedules

How many schedules of inspections have you completed for the installation? Often this will only be one.

How many schedules of test results have been completed? This will normally be one for each distribution/consumer unit.

Schedule of test results

This document is a generic document which can be used with either an electrical installation certificate or an electrical installation condition report.

Figure 6.3 shows the basic document; some certification bodies have certificates which are a little more comprehensive. It is very important that each box has an entry made in it, as this will prevent the document from being altered. The entry can be either a ✓, N/A or X.

Information required

DB reference number

Where there is more than one consumer unit/distribution board it is very important that they are identified. How is it identified: number, name or letter?

Location

Where is it?

Z_s at DB_W

For a single board this would be the Z_e, where the board is supplied by distribution circuit the value entered would be the Z_s for the distribution circuit. This value will then become the Z_e for any circuits fed by the board.

I_{pf} at DB (kA)

This is the fault current which has been measured at the board.

Correct supply polarity confirmed

Is the incoming supply correct?

Phase sequence confirmed (3 ph)

A rotating disc type or indicating lamp type of phase rotation type of tester should be used to check the correct phase sequence.

Details of circuits or installed equipment which may be vulnerable to damage when testing

Some electronic equipment can be damaged when tested, particularly when using 500v for insulation testing. Items such as PIRs, some RCDs and electronic controls would be recorded here.

Details of instruments used

Make and serial number should be recorded here.

When tested and by whom

Tester's name and test date are required.

Circuit details

This will include the following:

- Circuit number.
- Circuit description, Ring, lighting, cooker etc. As much detail as possible to ensure the circuit is easily identified.
- BS (EN) number of protective device, 60898, 3036 and others
- Type. Is it a B,C,D or perhaps a 1, 2 or 3?
- Rating. Operating current of the device.
- Reference method for the cables. Table 42A in Appendix 4 of BS 7671 will be of help here.
- Live mm^2. CSA of line and neutral conductors.
- CPC mm^2. CSA of circuit protective conductors.

Test results

Ring final circuit continuity

r_1, r_n and r_2 – this is the resistance of the line neutral and CPC each measured end to end.

Continuity Ω

$R_1 + R_2$ – this is the highest measurement recorded at a socket outlet when measured with the ring circuit line and CPC ends cross connected.

R_2 This is where a long lead has been used to verify that there is an earth present. A simple ✓ is all that is required here as it is very likely that there will be parallel paths present.

Insulation resistance

For an initial verification all conductors must be tested.

Live – Live is the measured value between line and N or between all phases and phases to N.

Live – Earth is measured between all live conductors and earth.

Polarity

Is the polarity correct at each circuit? Are the single pole protective devices and switches in the line conductor? Have ES lamp holders been correctly connected?

Z_s

This can be Z_e added to $R_1 + R_2$ or the result which has been measured live (the live test is preferable).

The calculated value of $Z_e + R_1 + R_2$ will not include parallel paths if carried out correctly whereas the measured Z_s will as this is a live test and all protective conductors must be connected for the test to be carried out safely. Therefore the measured Z_s should be the same as the $Z_e + R_1 + R_2$ value or even less if parallel paths are present. It should not be higher!

RCD

@ $I_{\Delta n}$ This is the tripping time at the rated tripping current of the RCD.

@ $5I_{\Delta n}$ This is the tripping time of the RCD at five times its rating, this will only be required for RCDs with a trip rating of up to and including 30mA.

Test button operation

This is a check on the mechanical operation of the RCD which should always be carried out last.

Remarks

This is the where you have to record anything which may be unusual, or which may be of use to anyone who may work on the installation at a later date.

Schedule of inspections

This document, which can be found in BS 7671 along with the schedule of test results (Figure 6.3), forms part of the electrical installation certificate; without this schedule the other certificates and reports are invalid.

Completion of this document involves the marking of boxes which must be marked using a ✓, X or NA after the inspection is made, and is useful if used also as a check list.

An X should never be entered onto a schedule which accompanies an EIC.

Electrical installation condition report

This document (Figure 6.4) is used to record the condition of an installation, in particular, is it safe to use? At the present time it is not a requirement for the person completing this report to be Part P compliant. It is important however that the person carrying out the inspection and test is competent.

The report must also include a schedule of test results (Figure 6.3) and a condition report schedule of inspection, which can be found in BS 7671.

A periodic inspection is carried out for many reasons, in particular:

- the due date
- client/customer request
- change of ownership
- change of use
- insurance purposes.
- to inspect the condition of the existing installation, prior to carrying out any alterations or additions.

The frequency of the periodic inspection and any testing which may be required is dependent on the type of installation, the environment and the type of use. BS 7671 Wiring Regulations refer to this as the 'Construction, utilisation and environment' and this can be found in Appendix 5 of BS 7671.

Guidance Note 3 for the inspecting and testing of electrical installations has a table of recommended frequencies for carrying out periodic inspections; this period depends on the type of installation.

The recommended frequencies are not cast in stone and it is the responsibility of the person carrying out the periodic inspection and test to decide on the period between tests. This decision should be based on the inspector's experience, what the installation is used for, how often it is used and the type of environment that surrounds the installation. These things and many others should be taken into account when setting the next test date.

It is important to remember that the date of the first inspection and test is set by the person responsible for the installation design. However circumstances change which could affect the installation, such as change of use or ownership.

Careful consideration must be given to the installation before the date of the next periodic inspection and test is set.

It is very important that the extent and limitations of the inspection and test is agreed with the person ordering the work before commencing work.

ELECTRICAL INSTALLATION CONDITION REPORT

Section A. Details of the person ordering the report
Name ....
Address...
....

Section B. Reason for this report being produced ....
....
Date on which the inspection was carried out...

Section C. Details of the installation which is the subject of this report.
Occupier..
Address...
....
Description of premises
Industrial ☐   Commercial ☐   Domestic ☐   Other (description of premises)...
Estimated age of wiring system................Years
Any evidence of additions or alterations   Yes ☐ No ☐   Non apparent ☐   If yes estimate of age ....Years
Are installation records available (reg 651.1) Yes ☐   No ☐ Last inspection date...................................

Section D. Extent and limitations of inspection and testing carried out
Extent of the electrical installation covered by this report...
....
Agreed limitations and reasons (Reg 653.2)..
....
Agreed with..
Operational limitations with reasons ....

The inspecting and testing detailed in this report and attached schedules have been carried out in accordance with BS 7671 : 2018 as amended to ....
Cables concealed within conduits and trunking. Inaccessible roof voids, under floors and other building voids have not been inspected unless specifically
agreed with all parties involved.

Section E. Summary of the condition of the installation
General condition of the installation in terms of electrical safety..
....
General assessment of the installation in terms of continued use   Delete as appropriate SATISFACTORY   /   UNSATISFACTORY
An unsatisfactory assessment indicates that a part or parts of the installation are dangerous (code C1) and/or parts of the installation are potentially
dangerous (code C2) and have been identified.

Section F. Recommendations
Where the overall assessment of the suitability for continued use is stated as UNSATISFACTORY, I/We recommend that any C1 (Dangerous) or C2 (potentially
dangerous) observations are acted upon as a matter of urgency.
Any items marked as further investigation required (code F1) should be investigated as soon as reasonably possible.
Any improvement recommendations (Code C3) should be considered.

Subject to the required remedial work being carried out I/We recommend that this installation is inspected and tested no later than ....Date

Section G. Declaration
I/We being the persons responsible for the inspection and testing of the electrical installation, particulars of which are described above, having exercised
reasonable skill and care when carrying out the inspection and testing, hereby declare that the information in this report, including the observations and the
attached schedules provide an accurate assessment of the condition of the installation taking into account the stated extent and limitations listed in section
D of this report.

Inspected and tested by:	**Report authorised for issue by:**
Name (Capitals)...	Name (Capitals)...
Signature..	Signature..
For and on behalf of...	For and on behalf of...
Position..	Position..
Address..	Address..
....	....
Date...	Date...

Section H : Schedule(s)
.... Schedule (s) of inspection and ....Schedules of test results are attached

The attached schedules form part of this document and the report is only valid when they are attached to it

Figure 6.4a Electrical installation condition report

Section I. Supply Characteristics and Earthing Arrangements

Earthing arrangements	Number and type of live conductors		Nature of Supply Parameters	Supply Protective Device
TT ☐ TN-S ☐ TN-C-S ☐ TN-C ☐ IT ☐	AC ☐ 1-phase,2 wire ☐ 2-phase, 3 wire ☐ 3-phase,3 wire ☐ 3-phase, 4 wire ☐	DC ☐ 2-wire ☐ 3-wire ☐ other ☐	Nominal Voltage, U /U$_o$ [(1)] ....V Nominal frequency, f[(1)] ....Hz Prospective fault current, I$_{pf}$[(2)] ....kA External loop impedance, Z$_s$[(2)] ....Ω Note (1) by enquiry (2) By measurement or enquiry	BS (EN) .... Type... Rated Current.......................................A

Other sources of supply (As attached information)

Section J: Particulars of The Installation Referred to on the report

Means of Earthing	Maximum Demand
Distributors facility ☐	Maximum Demand load...kVA/Amps (delete as necessary)

Installation earth electrode ☐	**Details of the installation earth electrode if installed**
	Type (e.g. Rods, tape, plate etc).. Location... Electrode resistance to earth....................Ω

Main Protective conductors

Earthing conductor	Material...................................... csa ....mm^2	Connection / continuity verified ☐
Main protective bonding conductors (to extraneous conductive parts)	Material.........................csa..............................mm^2	Connection / continuity verified ☐

To water installation pipes ☐	To gas installation pipes ☐	To oil installation pipes ☐	To structural steel ☐	To lighting protection ☐

To other ☐ ....

Main Switch / Fuse / Circuit Breaker / RCD

Location... .... BS(EN).. Number of poles....................................	Current rating...A Fuse / device rating...............................A Voltage rating V	**Where RCD is main switch** Rated residual operating current (I$_n$)...................mA Rated time delay...ms Measured operating time ....ms

Section K: Observations

Referring to the attached schedules of inspection and test results, and subject to the limitations specified in section D

No remedial action required ☐   The following observations are made ☐

Observation(s) Include schedule reference	Classification Code
....	....
....	....
....	....
....	....
....	....
....	....
....	....
....	....
....	....
....	....
....	....
....	....

One of the following codes, as appropriate, has been allocated to each of the observations made above to indicate to the person responsible for the installation the degree of urgency for remedial action.

C1 – Danger present, immediate remedial action required

C2 – Potentially dangerous – urgent remedial action required

C3 – Improvement recommended

F1 – Further investigation required without delay

Figure 6.4b Continued.

CONDITION REPORT INSPECTION SCHEDULE FOR DOMESTIC
AND SIMILAR PREMISES UP TO 100 A SUPPLY

OUTCOMES	Acceptable condition	✓	Unacceptable condition	State C1 or C2	Improvement recommended	State C3	Further investigation	F1	Not Verified	N/V	Limitation	LIM	Not Applicable	N/A
ITEM NO	**DESCRIPTION**								**OUTCOME** Use codes shown above and provide additional comments where required. Any item coded C1, C2, C3 and F1 must be recorded in section K of the condition report					

ITEM NO	DESCRIPTION	OUTCOME
1.0	External condition of intake equipment (visual check only)	
1.1	Service cable	
1.2	Service head	
1.3	Earthing arrangement	
1.4	Meter tails	
1.5	Metering equipment	
1.6	Isolator (where present)	
2.0	**Presence of suitable arrangements for other sources such as microgeneration (551.6, 551.7)**	
3.0	Earthing / Bonding Arrangements (411.3 Chap 54)	
3.1	Presence and condition of distributors earthing arrangement (542.1.2.1; 542.1.2.2)	
3.2	Presence and condition of the earth electrode if applicable (542.1.2.3)	
3.3	Presence of earthing and bonding labels in the correct locations (514.13.1)	
3.4	Confirmation of the earthing conductor size (542.3; 543.1.1)	
3.5	Accessibility and condition of the earthing conductor at the MET (543.3.2)	
3.6	Confirmation of any main protective bonding conductor sizes (544.1)	
3.7	Condition and accessibility of main protective conductor connection (543.3.2; 544.1.2)	
3.8	Accessibility and condition of other protective bonding connections (543.3.1 543.3.2)	
4.0	**Consumer unit (s) / Distribution boards**	
4.1	Adequacy of working space/access to consumer/distribution boards (132.12 : 513.1)	
4.2	Security of fixing (134.1.1)	
4.3	Condition of enclosures in terms of IP ratings etc (416.2)	
4.4	Condition of enclosures in terms of fire rating etc (421.1.201; 526.5)	
4.5	Enclosure not damaged so as to impair safety (651.2)	
4.6	Presence of main linked switch (462.1.201)	
4.7	Operation of main switch (functional check) (643.10)	
4.8	Manual operation of circuit breakers and RCDs to prove disconnection (643.10)	
4.9	Correct identification of circuits and protective devices (514.8.1; 514.9.1)	
4.10	Presence of RCD six-monthly test notice at/near consumer unit/distribution board (514.12.2)	
4.11	Presence of non-standard (mixed) cable colour warning notice at/near consumer unit/distribution board (514.14)	
4.12	Presence of alternative supply warning notice at/near consumer unit/distribution board (514.15)	
4.13	Presence of other required labelling (specify) (section 514)	
4.14	Compatibility of protective devices, bases and other components; correct type and rating (no signs of acceptable thermal damage, arcing and or overheating) (411.3.2; 411.4; 411.5; 411.6; sections 432,433)	
4.15	Single-pole switching or protective devices in the line conductor only (132.14.1; 530.3.3)	
4.16	Protection against mechanical damage where cables enter enclosures (132.14.1; 522.8.1; 522.8.5; 522.8.11)	
4.17	Protection against electromagnetic effects where cables enter ferrous enclosures (521.5.1)	
4.18	RCDs provided for fault protection- including RCBOs (411.3.3; 415.1)	
4.19	RCDs provided for additional protection including RCBOs (411.3.3; 415.1)	
4.20	Confirmation of information that SPD is functional (651.4)	
4.21	Confirmation that all conductor connections including connections to busbars, are correctly terminated and are tight and secure (526.1)	
4.22	Adequate arrangements where a generating set operates as a switched alternative to the public supply (551.6)	
4.23	Adequate arrangements where a generating set operates in parallel with a public supply (551.7)	

Figure 6.4c Continued.

OUTCOMES	Acceptable condition	✓	Unacceptable condition		State C1 or C2	Improvement recommended	State C3	Further investigation	F1	Not Verified	N/V	Limitation	LIM	Not Applicable	N/A
ITEM NO	**DESCRIPTION**									**OUTCOME** Use codes shown above and provide additional comments where required. Any item coded C1, C2, C3 and F1 must be recorded in section K of the condition report					

5.0	**Final Circuits**	
5.1	Identification of conductors (514.3.1)	
5.2	Cables correctly supported throughout their run (521.10.202; 522.8.5)	
5.3	Condition of insulation of live parts (416.1)	
5.4	Non-sheathed cables protected by enclosures in conduit, trunking or ducting (521.10.1)	
	• To include the integrity of the trunking and conduit systems	
5.5	Adequacy of cables for current carrying capacity with regard to the type and nature of the installation (section 523)	
5.6	Coordination between conductors and overload protective devices (433.1; 533.2.1)	
5.7	Adequacy of protective devices: type and rated current for fault protection (411.3)	
5.8	Presence and adequacy of circuit protective conductors (411.3.1; Section 543)	
5.9	Wiring system(s) appropriate for the type and nature of the installation and external influences (section 522)	
5.10	Concealed cables installed in prescribed zones (see section D, Extent and limitations) (5226.204)	
5.11	Cables concealed under floors, above ceilings or in walls/partitions, adequately protected against damage (see section D, Extent and limitations) (5226.204)	
5.12	Provision of additional requirements for protection by RCD not exceeding 30mA	
	• For all socket outlets of rating 32A or less, unless an exception is permitted (411.3.3)	
	• Supply of mobile equipment not exceeding 32A rating for use outdoors (411.3.3)	
	• For cables installed in walls at depth of less than 50 mm (522.6.202; 522.6.203)	
	• For cables concealed in walls/partitions containing metal parts regardless of depth (522.6.203)	
	• Final circuits supplying luminaires within domestic (household) premises (411.3.4)	
5.13	Provision of fire barriers, sealing arrangements and protection against thermal effects (section 527)	
5.14	Band II cables segregated / separated from Band I cables (528.1)	
5.15	Cables segregated / separated from communications cables (528.2)	
5.16	Cables segregated / separated from non-electrical services (528.3)	
5.17	Termination of cables at enclosures- indicate the extent of sampling in section D of the report (section 526)	
	• Connection soundly made and under no strain (526.6)	
	• No basic insulation of a conductor visible outside of an enclosure (526.8)	
	• Connection of live conductors adequately enclosed (526.5)	
	• Adequately connected at the point of entry into an enclosure(bushes, glands etc) (522.8.5)	
5.18	Condition of accessories including socket-outlets, switches and joint boxes (651.2(v))	
5.19	Suitability of accessories for external influences (512.2)	
5.20	Adequacy of working space/ accessibility of equipment (132.12; 513.1)	
5.21	Single-pole switching or protective devices in the line conductor only (132.14.1; 530.3.3)	

6.0	**Location(s) containing a bath or a shower**	
6.1	Additional protection for all low voltage circuits by an RCD not exceeding 30mA (701.411.3.3)	
6.2	Where used as a protection measure, the requirements of SELV and PELV are met (701.414.4.5)	
6.3	Shaver sockets comply with BS EN 61558-2-5 formerly BS 3535 (701.512.3)	
6.4	Presence of supplementary bonding conductors, unless not required by BS 7671:2018 (701.415.2)	
6.5	Low voltage socket-outlets sited at least 3 m from zone 1 (701.512.3)	
6.6	Suitability of equipment for external influences for the installed location in terms of IP rating (701.512.2)	
6.7	Suitability of accessories and control gear etc., for a particular zone (701.512.3)	
6.8	Suitability of current using equipment for the particular position within the location (701.55)	

7.0	**Other part 7 special installations or locations**	
7.1	List all other special locations or installations present, if any (record separately the results of any inspections applied)	

Inspected by;

Name (Capitals)...   **Signature**..   **Date**..

Figure 6.4d Continued.

Before the extent and limitation can be agreed, discussion between all parties involved must take place. The client will know why they want the inspection carried out and the person who is carrying out the inspection and test should have the technical knowledge and experience to give the correct guidance.

Past test results, electrical installation or periodic inspection reports, installation condition reports, fuse charts, etc. must be made available to the person carrying out the inspection.

If these are not available, then a survey of the installation must be carried out to ensure that the installation is safe to test and to prepare the required paperwork, such as fuse charts.

Whilst carrying out a periodic inspection it is not a requirement to take the installation apart. It should be carried out with the minimum of intrusion; disconnection should only be carried out where it is impossible to carry out the required test in any other way.

For example if an insulation resistance test is required on a lighting circuit with fluorescent lighting connected to it, the simple method would be to open the switch supplying the fluorescent fitting before testing between the live conductors and close the switch when conducting the test between live conductors and earth. It is not a requirement to disconnect the fitting (see insulation resistance testing).

Completing the form

As with a minor works or an electrical installation certificate the information required on an electrical installation condition report will vary depending on where the report is obtained from.

It will include the following:

Details of the client
Name and address of the person who has ordered the work. This is not necessarily the installation address.

Reason for producing the report
Is it the due date, insurance purposes etc.

Details of the installation

* Occupier: this may not be the owner.
* Installation address: this is the installation which is being inspected.
* Description of the premises: domestic, commercial, industrial, other (include brief description).
* Estimated age of wiring system in years.

- Evidence of additions/alterations. If yes, estimate the age in years.
- Installation records available? (Regulation 621.1)
- Date of last inspection (date).

Extent and limitation of inspecting and testing

- Extent of the electrical installation covered by this report: what does the inspection cover, is it the complete installation or just a part of it?
- Agreed limitations including the reasons (see Regulation 653.2): are there any areas that are not being inspected, are there circuits which cannot be isolated or tested and why?
- Agreed with: who have the limitations been agreed with?
- Operational limitations including the reasons: areas not to entered or access not permitted.

  The inspection and testing detailed in this report and accompanying schedules have been carried out in accordance with BS 7671:2018 amended to …

It should be noted that cables concealed within trunking and conduits, under floors, in roof spaces, and generally within the fabric of the building or underground, have *not* been inspected unless specifically agreed between the client and inspector prior to the inspection. You can only report on what you can see. On occasion a judgement may have to be made but this will require a greater level of experience and knowledge than that required for an initial verification.

Summary on the condition of the installation

General condition of the installation

This section is where you would record any concerns which you have about the electrical safety of the installation. It is not asking if the installation is untidy or if the sockets and switches are straight.

Overall assessment of the installation in terms of its suitability for continued use

Satisfactory or unsatisfactory should be entered here. If unsatisfactory is entered it will indicate that the installation is either in a dangerous state (C1) (example would be exposed conductors) or is potentially dangerous (C2)(example would be high Z_s values).

An explanation of what is making the installation unsatisfactory is required; entries such as fuse board change required or rewire required are unsuitable as reasons should be given. Often these explanations will need to be entered onto a separate sheet.

Recommendations

Where the overall assessment of the suitability of the installation for continued use above is stated as UNSATISFACTORY, I/We recommend that any observations classified as 'Danger present' (code C1) or 'Potentially dangerous' (code C2) are acted upon as a matter of urgency.

Investigation without delay is recommended for observations identified as 'further investigation required'.

Observations classified as 'Improvement recommended' (code C3) should be given due consideration.

This section is where you need to record the actions which should be taken to make the installation safe for continued use.

Subject to the necessary remedial action being taken, I/We recommend that the installation is further inspected and tested by …. Date:

This is where the recommended date of the next inspection is entered. This will depend on the general condition, type of use and the environment.

Declaration

I/We, being the person(s) responsible for the inspection and testing of the electrical installation (as indicated by my/our signatures below), particulars of which are described above, having exercised reasonable skill and care when carrying out the inspection and testing, hereby declare that the information in this report, including the observations and the attached schedules, provides an accurate assessment of the condition of the electrical installation taking into account the stated extent and limitations in section D of this report.

This is the section where you sign to indicate that you have carried out the inspection correctly using the required skills and competences.

Schedules

Some installations will have more than one schedule, particularly installations which have more than one consumer unit or distribution board.

Supply characteristics and earthing arrangements

Earthing

Is it TT, TNS or TNCS?

Number and type of live conductors

Usually 1 phase 2 wire or 3 phase 4 wire (only live conductors).

Nature of supply parameters

This can be gained by enquiry or measurement.

U is line to line or line to neutral

U_o *is line to earth.*

Do not record measured values. If three phase or three phase and neutral then the value will be U 400 v. If single phase then the U_0 will be 230v.

Frequency

This will normally be 50 Hz although on some special installations this may change.

Prospective fault current (I_{pf})

Prospective fault current is the highest current that could flow within the installation between live conductors, or live conductors and earth. This should be measured or obtained by enquiry. If it is measured, remember that on a three phase system the value between phase and neutral must be doubled.

External earth loop impedance (Z_e)

This should be measured between the phase and earth on the live side of the main switch with the earthing conductor disconnected. (Remember to isolate installation first.) If measurement is not possible then it can be obtained by enquiry to the supply provider.

Supply protective device

- *BS type*. Can normally be found printed on the service head.
- *Nominal current rating*. Can normally be found printed on the service head.
- *Short circuit capacity.* This will depend on the type, but if in doubt reference should be made to Table 7.2.7(i) in the *On-site Guide*.
- *Main switch or circuit breaker*. Type, this is normally printed on it but reference can be made to Appendix 2 of BS 7671 if required.

- *Number of poles*. Does the switch break all live conductors when opened, or is it single pole only?
- *Supply conductor material and size*. This refers to the meter tails.
- Voltage rating. This will usually be printed on the device.
- Current rating. This will usually be printed on the device.
- *RCD operating current $I_{\Delta n}$*. This is the trip rating of the RCD and should only be recorded if the RCD is used as a main switch.
- *RCD operating time at $I_{\Delta n}$*. Only to be recorded if the RCD is used as a main switch.
- *Type*. Usually this will be a BS 88 or BS 1361 cartridge fuse and it will normally be marked on the supply cut out. If it is not then it should be found by enquiry to the supply provider.
- *Rated current*. This is the current which the protective device is rated at, it will normally be found printed on the fuse carrier.

Particulars of the installation referred to in the certificate

Means of earthing

- Distributors facility or earth electrode?
- Type of earth electrode.
- Electrode resistance. Usually measured as Z_e.
- Location. Where is the earth electrode?
- Method of measurement. Has an earth fault loop tester or an earth electrode tester been used to carry out the test?

To carry out this test correctly the earthing conductor should be disconnected to avoid the introduction of parallel paths. This will of course require isolation of the installation; in some instances this may not be practical or possible for various reasons. If isolation is not possible the measurement should still be carried out to prove that the installation has an earth. The measured value of Z_e should be equal to or less than any value for Z_e documented on previous test certificates. If the measurement is higher than those recorded before then further investigation will be required.

The higher measurement could be corrosion, a loose connection or damage.

If the means of earthing is by an earth electrode the soil conditions may have changed, this would be considered normal providing that the measured value is less than 200 Ω and the system is protected by a residual current device.

Details of earth electrode

If the system has an earth electrode, what type is it? Where is it? What is its resistance? (This is usually measured with an earth loop impedance tester, using the same method as for Z_e.)

Main protective conductors

Earthing conductor

What is it made of and what size is it? Has the connection been verified?

Main protective bonding conductors

What is it made of and what size is it? Has the connection been verified?

Is the bonding connected to the incoming water and gas service?

Other elements

Are there any other parts of the installation which have protective bonding connected? This could be incoming oil lines, central heating etc.

Main switch or circuit breaker

- *What type of switch is it?* BS, BS EN or IEC
- *Number of poles.* Single pole, double pole, triple pole or triple pole and neutral.
- *Voltage and current rating.* Whatever is marked on it.
- *Location.* Where is it?
- *Fuse rating or setting.* Where the main switch is a switched fuse or circuit breaker.
- *Rated residual operating current $I_{\Delta n}$.* If an RCD is fitted as a *main switch*, the operating current $I_{\Delta n}$ and the operating time at $I_{\Delta n}$ must be recorded.

Observations

Referring to the attached schedules of inspection and test results, and subject to the limitations specified at the Extent and limitations of inspection and testing section

In this section you must indicate whether or any remedial actions are required, where remedial actions are required, the observations box must be ticked and a list of observations, complete with the classification code must be recorded:

C1 – Danger present. Risk of injury. Immediate remedial action required

C2 – Potentially dangerous. Urgent remedial action required

C3 – Improvement recommended.

C1 This code is given when there is an immediate risk of danger. This could be a bare live conductor or perhaps a blank or cover missing to an electrical enclosure which would allow someone to touch a live part.

C2 This code is given where a fault would need to occur which would then cause the installation to be dangerous. An example would be if there was not an earth present. In this instance it would not present a problem until the installation was subjected to an earth fault.

The electrical safety council provide some very good best practice guides which are available free of charge at: www.esc.org.uk/industry/industry-guidance/best-practice-guides/

Guide number 4 provides some very good information on the completion of condition reports.

Condition report inspection schedule

This document, which can be found in BS 7671, is really a check list for the installation; all boxes must have an entry made in them although for some it may just be N/A.

A schedule of test results is also required and this is the same document as is used with the electrical installation certificate.

CHAPTER 7

Correct selection of protective devices

Protective devices are mentioned throughout this book – this chapter brings all of the information together for reference.

When carrying out an inspection and test on any electrical installation it is important to ensure that the correct size and type of device has been installed.

To do this we must have a good knowledge about the selection of protective devices and the type of circuits that they are protecting.

Why are they installed?

Protective devices are installed to protect the cable of the circuit from damage which could be caused by *overload, overcurrent* and *fault current*.

The definition for overload given in Part 2 of BS 7671:

> *Overcurrent occurring in a circuit which is electrically sound.*

This is when the circuit is installed correctly and the equipment connected to it is drawing too much current.

For instance it could be that an electric motor connected to the circuit is used on too heavy a load. This would overload the circuit and provided that the correct size of protective device was installed, the device will operate and interrupt the supply preventing the cable from overloading.

If additional luminaries were installed on an existing circuit which was already fully loaded, the protective device should operate and protect the cable of the circuit.

Overcurrent is a current flow in a circuit which is greater than the rated current carrying capacity of the cables.

This would normally be due to a fault on the circuit or incorrect cable selection.

Example 1

A 20 amp cable is protected by a 32 amp MCB. If a load of 25A was connected to this circuit the cable would overheat and the device would continue to allow current to flow. This could damage the cable.

Fault current is a current which is flowing in a circuit due to a fault.

Example 2

A nail is driven through a cable causing an earth fault or a short circuit fault. This would cause a very high current to flow through the circuit which must be interrupted before the conductors reach a temperature that could damage the insulation or even the conductors.

So what are we looking for with regard to protective devices during an inspection?

What type of device is it? Is it a fuse or circuit breaker?

A fuse has an element which melts when too much current is passed through it, whether by overload or fault current.

Fuses in common use are:

- BS 3036 semi rewirable fuse
- BS 88 cartridge fuse
- BS 1361 cartridge fuse.

A circuit breaker is really two devices in one unit. The overload part of the device is a thermal bi metal strip, which heats up when a current of a higher value than the nominal current rating (I_n) of the device passes through it.

Also incorporated within the device is a magnetic trip which operates and causes the device to trip when a fault current flows through it.

For the device to operate correctly it must operate within 0.1 seconds. The current which is has to flow to operate the device in the required time has the symbol (I_a).

Circuit breakers in common use are:

- BS 3871 types 1, 2 and 3
- BS EN types B, C and D (A is not used, this is to avoid confusion with Amp).

Is the device being used for protection against indirect contact?

In most instances this will be the case.

What type of circuit is the device protecting? Is it supplying fixed equipment only, or could it supply hand held equipment?

For any circuit rated up to and including 63A for socket outlets and 32A for fixed equipment, the device must operate on fault current within 0.4 of a second. Exceptions to this would be a distribution circuit or circuits supplied from a TT system. See BS 7671 Regulations 411.3.2.2 and 413.3.2.3.

When using circuit breakers to BS 3871 and BS EN 60898 these times can be disregarded. Providing the correct Z_s values are met they will operate in 0.1 seconds or less.

If it is a circuit breaker, is it the correct type?

Table 7.2.7(ii) of the *On-site Guide* provides a good reference for this.

Types 1 and B should be used on circuits having only resistive loads.

(Have you ever plugged in your 110v site transformer and found that it operated the circuit breaker? If you have, it will be because it was a type 1 or B.)

Types 2, C and 3 should be used for inductive loads such as fluorescent lighting, small electric motors and other circuits where surges could occur.

Types 4 and D should be used on circuits supplying large transformers or any circuits where high inrush currents could occur.

Will the device be able to safely interrupt the prospective fault current which could flow in the event of a fault?

Table 7.2.7(i) of the *On-site Guide* or manufacturers' literature will provide information on the rated short circuit capacity of protective devices.

Is the device correctly coordinated with the load and the cable?

Correct coordination is defined as:

Current carrying capacity of the cable under its installed conditions must be equal to or greater than the rated current of the protective device (I_z).

The rated current carrying capacity of the protective device (I_n) must be equal to or greater than the design current of the load (I_b).

In short $I_z > I_n > I_b$ (Appendix 4 item 4 BS 7671 or Appendix B of the *On-site Guide*).

Additional information regarding circuit breakers

Overload current

The symbol for the current required to cause a protective device to operate within the required time on overload is (I_2).

Circuit breakers with nominal ratings up to 60 amps must operate within 1 hour at 1.45 × their nominal rating (I_n).

Circuit breakers with nominal ratings above 60 amps must operate within 2 hours at 1.45 × their rating (I_n).

At 2.55 times the nominal rating (I_n) circuit breakers up to 32 amps must operate within 1 minute and circuit breakers above 32 amps must operate within 2 minutes.

They must *not* trip within 1 hour at up to 1.13 × their nominal rating (I_n).

Maximum earth fault loop impedance values (Z_s) for circuit breakers

These values can be found in Table 41.3 of BS 7671.

Because these devices are required to operate within 0.1 of a second, they will satisfy the requirements of BS 7671 with regard to disconnection times in all areas. Therefore the Z_s values for these devices are the same wherever they are to be used. (This only applies to circuit breakers even in special locations.)

Calculation of the maximum Z_s of circuit breakers

It is often useful to be able to calculate the maximum Z_s value for circuit breakers without the use of tables. This is quite a simple process for BS 3871 and BS EN 60898 devices.

Let's use a 20 amp BS EN 60898 device as an example. Table 7.2.7(ii) in the *On-site Guide* shows that a type B device must operate within a window of 3 to 5 times its rating. As electricians we always look at the worst case scenario, therefore we must assume that the device will not operate until a current equal to 5 times its rating flows through it (I_a).

For a 20 amp type B device this will be $5 \times 20a = 100a$.

If we now use a supply voltage of 230 volts, which is the assumed open circuit voltage (U_{oc}) of the supply (Appendix 3 BS 7671), Ohm's law can be used to calculate the maximum Z_s.

$$Z_s = \frac{U_o}{100} \qquad Z_s = \frac{230 \times 0.95}{100} = 2.185\Omega$$

Just as a check, if we now look in Table 41.3 of BS 7671 we will see that the value Z_s for a 20 amp type B device is 2.19Ω

Now let's use the same procedure for a 20 amp type C device. Table 7.2.7(ii) in the *On-site Guide* shows us that a type C device must operate at a maximum of 10 times its rating (I_n).

$$10 \times 20 = 200A$$

$$\frac{230 \times 0.95}{200} = 1.09\Omega$$

If we check again in Table 41.3 we will see that the maximum Z_s for a 20 amp type C device is 1.09Ω.

A type D circuit breaker with a nominal operating current (I_n) must operate at a maximum of 20 times its rating.

$$20 \times 20 = 400A$$

$$\frac{230 \times 0.95}{400} = 0.546\Omega$$

Again if we check in Table 41.3 we will see that the Z_s value is 0.55Ω.

We can see that the maximum Z_s values for a type C are 50 per cent of the Z_s value of a type B device, and that the Z_s value for a type D are 50 per cent of the Z_s value of a type C device.

Comparing maximum Z_s and measured Z_s

Unfortunately we cannot compare this value directly to any measured Z_s values that we have because the values given in BS 7671 for Z_s are for when the circuit conductors are at their operating temperature (generally 70°C).

We can however use a simple calculation which is called the rule of thumb (Appendix 14 of BS 7671). This calculation will allow us to compare our measured values with the values from BS 7671.

The values given in the tables in Part 4 of BS 7671 are the worst case values. In these types of calculations we must always use the worst case values to ensure a safe installation.

Table 7.1 shows that types D, 3 and 4 will have very low maximum permitted Z_s values which will often result in the use of an RCD.

Table 7.1 Circuit breaker application

Circuit breaker type	Worst case tripping current	Typical uses
BS EN 60898 B	5 times its rating	General purpose with very small surge currents (small amount of fluorescent lighting), mainly domestic.
C	10 times its rating	Inductive loads. Generally commercial or industrial where higher switching surges would be found (large amounts of fluorescent lighting or motor starting)
D	20 times its rating	Only to be used where equipment with very high inrush currents would be found
BS 3871 1	4 times its rating	As for type B
2	7 times its rating	As for type C
3	10 times its rating	As for type C but slightly higher inrush currents
4	50 times its rating	As for type D

Example 3

Let's assume that we have a circuit protected by a 32A BS EN 60898 type B device.

The measured value of Z_s is 0.98Ω.

Following the procedure described previously:

$$5 \times 32 = 160\Omega$$

$$\frac{230 \times 0.95}{160} = 1.365$$

Maximum Z_s at 70°C for the circuit is 1.37Ω.

To find the corrected value we can multiply 1.37×0.8

$$1.37 \times 0.8 = 1.096 \ (1.1)$$

1.1Ω now becomes our maximum value and we can compare our measured value directly to it without having to consider the ambient temperature or the conductor operating temperature.

Our measured value must be less than the corrected maximum value. In this case it is and the 32A type B device would be safe to use.

This type of calculation must be understood by any student studying for the City and Guilds 2394/5 inspecting and testing courses, as well as the 2396 design and verification course.

Testing transformers

It is a requirement to test isolation and SELV transformers to ensure the users' safety. It is also useful to be able to test them to ensure that they are working correctly.

Video footage is also available on the companion website for this book.

Step up or down double wound transformer

Use a low resistance ohm meter to test between to primary *(cable that connects to the main supply)* side, the resistance should be quite high. This will of course depend on the size of the transformer. It may be that the resistance is so high a multimeter set on its highest resistance value will have to be used. If this is the case then set the instrument to the highest value possible and turn it down until a reading is given. If the winding is open circuit then the transformer is faulty. Repeat this test on the secondary winding.

Now join the ends of the primary winding together and join the ends of the secondary winding together.

Use an insulation resistance meter set on 500v d.c. to test between the joined ends.

Now test between the joined ends and earth.

The maximum insulation value permissible in both cases is 1MΩ. If the resistance is less, then the transformer is faulty.

Isolation transformer

Carry out the test in the same manner as the double wound transformer. The minimum acceptable value for insulation resistance is 1MΩ.

Separated extra low voltage transformers (SELV and PELV)

These transformers are tested using the same procedure as for the step up or down transformer. The insulation resistance test values are different for this test. If the SELV or PELV circuits from the secondary side of the transformer are being tested, then the test voltage must be 250v d.c. and the maximum resistance value is 0.5MΩ although this would be considered a very low value and any value below 5MΩ must be investigated.

For a test between the actual transformer windings the test voltage is increased to 500 volts d.c. The minimum insulation resistance value is 1MΩ although any value below 5MΩ should be investigated.

FELV

FELV transformers must be tested using the same method as for SELV transformers, however the text voltages and results must meet the requirements for low voltage circuits.

Video footage is also available on the companion website for this book.

Testing a three phase induction motor

There are many types of three phase motor but by far the most common is the induction motor. It is quite useful to be able to test them for serviceability.

Before carrying out electrical tests it is a good idea to ensure that the rotor turns freely. This may involve disconnecting any mechanical loads. The rotor should rotate easily and you should not be able to hear any rumbling from the motor bearings.

Next if the motor has a fan on the outside of it, check that it is clear of any debris which may have been sucked in to it; also check that any air vents into the motor are not blocked.

Generally if the motor windings are burnt out there will be an unmistakable smell of burnt varnish, however it is still a good idea to test the windings as the smell could be from the motor being overloaded.

Three phase motors are made up of three separate windings. In the terminal box there will be six terminals as each motor winding will have two ends. The ends of the motor windings will usually be identified as W1, W2, U1, U2, V1, V2.

The first part of the test is carried out using a low resistance ohm meter. Test each winding end to end W1 to W2, U1 to U2 and V1 to V2. The resistance of each winding should be approximately the same and the resistance value will depend on the size of the motor. If the resistance values are different then the motor will not be electrically balanced and it should be sent for rewinding.

If resistance values are the same then the next test is carried out using an insulation resistance tester. Join W1 and W2 together, U1 and U2 together and V1 and V2 together. Carry out an insulation resistance test between the joined ends i.e: W to U then W to V and then between U and V, then repeat the test between joined ends and the case, or the earthing terminal of the motor *(these tests can be in any order to suit you)*.

Providing the insulation resistance is 2MΩ or greater then the motor is fine. If the insulation resistance is above 0.5MΩ then it could just be dampness and it is often a good idea to run the motor for a while and then carry out the insulation test again as the motor may dry out with use.

To reconnect the motor windings in star, join W2, U2 and V2 together and connect the three phase motor supply to W1, U1 and V1. If the motor rotates in the wrong direction swop two of the phases of the motor supply.

To reconnect the motor windings in delta, join W1 to U2, U1 to V2 and V1 to W2 and then connect the three phase motor supply one to each of the joined ends. If the motor rotates in the wrong direction, swop two phases of the motor supply.

Polarisation index testing of an electric motor

Video footage is also available on the companion website for this book.

Polarisation index testing of an electric motor

Polarisation testing is carried out on a motor to determine 4 things:

- Detect any moisture present in a motor
- Any gradual deterioration of the insulation around the motor windings.
- The electrical condition of the motor
- Suitability for continued use.

This test is considered one of the best for judging the performance of a motor.

The life of a motor is dependent on the condition of the winding insulation. There are two tests which are used to identify the condition of a motor's windings. One of these tests is an insulation resistance test and the other is known as a polar index test.

Before the testing of a motor is carried out it is important to prepare the motor.

- Disconnect the motor
- If it is three phase use a low resistance ohm meter to test each of the windings end to end. You will have 6 winding ends. Usually they are marked as W1 – W2, U1 – U2, V1 – V2. Each letter denotes the separate ends of each winding.
- If they are not marked just mark them yourself and make a diagram to show which ends are connected where. This is to ensure that you are putting it back correctly.
- Test each winding end to end. The values should be very close, say, within 5 per cent.

If for some reason you muddle up the ends of the windings, don't worry – it is a simple job to sort them out. All that is required is a low resistance ohm meter.

Connect one test lead to one end of a winding, now touch the other test lead on the ends of the other windings. When you have continuity you will have found the other end of the winding to which the test lead is connected. Mark the winding to show that the ends are a pair.

Now connect the test lead to one of the remaining four ends and touch the other lead to each of the other three ends until you have continuity; mark these two leads as a pair.

The remaining ends should also be a pair but it is always worth checking.

Providing the resistance values of the windings are very similar we can now carry out the insulation resistance test between each of the windings.

- Join the ends of one winding together (say, W1 W2) and test between them and the other windings either together or separately.
- Carry out this test between all sets of windings.
- The minimum acceptable value for this test is 1MΩ, although hopefully it will be much higher.

Once this test has been completed it is important that all of the ends of the windings are shorted out to the case of the motor. This will ensure that any stored energy due to capacitance is drained from the windings.

Using an insulation tester set at 500volts, test between all motor windings joined and the case of the motor. Keep the test going for 1 minute and then record the result.

If at this point a low value is recorded it may be just because the motor has been standing idle and is possibly damp. In these situations it is often a good idea to connect the motor and run it for a few hours or even just put it in a warm place for a few days. Either of these options should dry it out.

Now repeat the test for a period of 10 minutes and record the reading.

Next, divide the 10 minute resistance value by the 1 minute resistance value. This will indicate the condition of the motor windings.

Insulation condition (PI):

Dangerous – Less than 1
Poor – Less than 1.5
Questionable – 1.5–2
Fair – 2–3
Good – 3–4
Excellent – Greater than 4

CHAPTER 8

Test equipment

It is important that the test equipment you choose is suitable for your needs. Some electricians prefer to use individual items of equipment for each test while others like to use multi-function instruments.

Any test instruments used for testing in areas such as petrol filling stations, or areas where there are banks of storage batteries etc. in fact anywhere there is a risk of explosion, must be intrinsically safe and suitable for the purposed use.

Most electricians are aware that electrical test instruments must comply with BS standards; however, for students studying for Part P and City and Guilds 2391/2400 exams, an understanding of the basic operational requirement for the most common types of test instrument is very important.

Whichever instrument you choose it must suitable for the use to which it's to be put and be manufactured to the required British Standard. It is also vital that you fully understand how to operate it before you start testing.

Instruments required

Low resistance ohm meter

This is used to measure the resistance and verify the continuity of conductors. This instrument must produce a test voltage of between 4 and 24 volts and a current of not less than 200mA. The range required is 0.2Ω to 2Ω with a resolution of at least 0.01Ω (in other words the range must go up or down in steps of 0.01Ω) although most modern instruments are self-ranging and will measure higher values if required.

Insulation resistance tester

This is used to measure the insulation resistance between live conductors and live conductors and earth. This test is best described as a pressure test applied to the conductor insulation.

The instrument must deliver a current of 1mA on a resistance of 0.5MΩ. At 250 volts d.c. for extra low voltage circuits, 500 volts d.c. for low voltage circuits up to 500 volts a.c. and 1000 volts d.c. for circuits between 500 and 1000 volts d.c.

This instrument is sometimes called a high resistance tester as it measures values in megohms.

Earth fault loop impedance tester

This instrument allows a current of up to 25A to flow around the earth fault loop path; it measures the current flow and by doing so can calculate the resistance of the earth fault loop path. The values given are in ohms. The test instrument should automatically cut off the test current after 40ms; this is to reduce the risk of shock hazard.

As a current of 25A will trip RCDs and some smaller circuit breakers, it is useful to have an instrument that can carry out low current testing where required. Use of this type of instrument will avoid the tripping of devices during testing.

Prospective short circuit current test instrument

This instrument measures the current that would flow between live conductors in the event of a short circuit. It is usually incorporated in the earth loop impedance tester and normally gives a value in kA. Some instruments give the value in ohms which then needs to be converted to amps by using Ohm's law (*use 230 volts*).

Most instruments will measure the value between phase and neutral and not between phases. To find the value between phases it is simply a matter of doubling the phase to neutral value.

Earth electrode resistance tester

This is normally a battery operated 3 or 4 terminal instrument with a current and potential spike. The values given are in ohms and the instrument instructions should be fully understood before using it. The instrument would be used where low and very accurate earth electrode resistance values are required such as for generators or transformers.

Residual current device tester

This instrument measures the tripping times of RCDs in seconds or milliseconds.

Phase rotation

This instrument is used to ensure the correct phase rotation of three phase supplies.

This type of tester will show the phase rotation by using a rotating disc or indicator lamps and should be capable of continued operation.

Phase rotation, phase sequence test

Video footage is also available on the companion website for this book.

Thermographic equipment

The use of this type of equipment is not recognised by BS 7671 (yet) but can be very useful in the detection of overloaded circuits and loose connections.

Calibration of test instruments

To carry out any kind of test properly your instruments have to be accurate; if they were not then the whole point of carrying out the test would be lost.

It is not a requirement to have instruments calibrated on an annual basis. However a record must be kept to show that the instruments are regularly checked for accuracy.

Instrument accuracy can be tested using various methods. For an earth loop impedance tester all that is required is a dedicated socket outlet. Use your earth fault loop impedance instrument to measure the value of the socket outlet. This value can then be used as a reference to test the accuracy of the instrument at a later date and you can also test any other earth fault loop instrument on the dedicated outlet to check its accuracy. The loop impedance values of the socket outlet should not change.

For an insulation resistance tester or a low resistance ohm meter the accuracy can be checked quite simply by using various values of resistors. The instruments could even be checked against values given by another instrument. When testing against another instrument

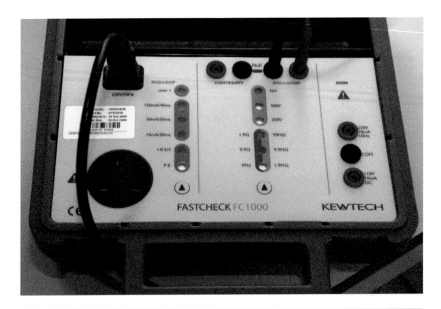

Figure 8.1 Test box

if the values given are not the same it will indicate that one of the instruments is inaccurate and further investigation using resistors should be undertaken.

An RCD test instrument accuracy is a little more difficult to check and often the best way is to check it against another instrument. However, if you do check it in this way do not expect exactly the same values as the trip time could be slightly different each time you test it due to the instrument increasing slightly in temperature.

It is also possible to purchase a test box which will check the accuracy of all electrical test instruments (Figure 8.1). When it is found that an instrument is not accurate, then it must be returned to the manufacturer or specialist for recalibration, it is not a job that can normally be carried out by the owner of the instrument.

Record keeping for the testing of the accuracy is quite a simple but very important process.

A record showing the instrument model, serial number and the date of the test along with the recorded values is all that is required and will satisfy most regulatory bodies. Records can be kept in a ledger, on a computer or calibration registers can be purchased to make life a little easier (Figure 8.2).

If for any reason your instrument does require recalibration it should be returned to the instrument manufacturer or a calibration specialist.

Megger. RECORD CARD   COMPANY NAME

| TEST BOX MTB7671 | SERIAL No. CALIBRATION DUE: | | | | | TESTED INSTRUMENT MODEL: SERIAL No. |

TEST DATE	VOLTS (V)	CONTINUITY I TEST (√)	V TEST (√)	0.5Ω (Ω)	5Ω (Ω)	INSULATION 250V 0.25MΩ (MΩ)	500V 0.5MΩ (MΩ)	1kV 1MΩ (MΩ)	9MΩ (MΩ)	90MΩ (MΩ)	TESTED BY
23/1/2007	242	√	√	0.49	5.1	0.25 v √ I √	0.51 v √ I √	1.05 v √ I √	9.1 v √ I √	89.8 v √ I √	X Ample
1						v I	v I	v I	v I	v I	
2						v I	v I	v I	v I	v I	
3						v I	v I	v I	v I	v I	
4						v I	v I	v I	v I	v I	
5						v I	v I	v I	v I	v I	
6						v I	v I	v I	v I	v I	
7						v I	v I	v I	v I	v I	
8						v I	v I	v I	v I	v I	

www.megger.com   Megger Record Set 6173-032 Ed3

Megger. RECORD CARD   COMPANY NAME

| TEST BOX MTB7671 | SERIAL No. CALIBRATION DUE: | | | | | TESTED INSTRUMENT MODEL: SERIAL No. |

TEST DATE	VOLTS (V)	LOOP (Ω)	LOOP +1Ω (Ω)	LOOP +180Ω (Ω)	PFC (kA)	RCD POL. 0/180°	RCD 10mA ½I 5mA (ms)	I 10mA (ms)	5I 50mA (ms)	RCD 30mA ½I 15mA (ms)	I 30mA (ms)	5I 150mA (ms)	RCD 100mA ½I 50mA (ms)	I 100mA (ms)	5I 500mA (ms)	TESTED BY
23/1/2007	239	0.12	1.12	180.1	2.02	0°	>1999	40	10	>1999	40	10	>1999	40	10	X Ample
						180°		50	20		50	20		50	20	
1						0°										
						180°										
2						0°										
						180°										
3						0°										
						180°										
4						0°										
						180°										
5						0°										
						180°										
6						0°										
						180°										
7						0°										
						180°										
8						0°										
						180°										

Figure 8.2 Calibration register

Volt stick

Most electricians are well aware that volt sticks should not be used for the testing of live conductors, particularly while carrying out safe isolation procedures.

They are very useful though for giving an indication as to whether or not a piece of equipment is earthed.

Video footage is also available on the companion website for this book.

If you pass a volt stick over the metal case of a piece of equipment which is earthed it will keep flashing as just as it would if the equipment were isolated. If, however, the volt stick lights continuously it is a very good indication that the equipment is not earthed and further investigation must carried out.

Whenever a volt stick remains lit it is usually a very good warning that care should be taken.

CHAPTER 9

Electric shock

Electric shock is caused by current flowing through a body. A very small amount, between 50 and 80mA, is considered to be lethal to most human beings, although this would of course depend on the person's health and other circumstances. In livestock the lethal current would be considerably less.

The electrical regulations are set out to provide for the safety of persons and livestock. Electric shock is one risk of injury; Chapter 13 of BS 7671, Regulation 131.1 lists the other risks as:

- Excessive temperatures likely to cause burns, fire and other injurious effects.
- Mechanical movement of electrically actuated equipment, in so far as such injury is intended to be prevented by electrical emergency switching or by switching for mechanical maintenance of non-electrical parts of such equipment.
- Explosion.

Regulation 130.1 tells us that persons and livestock shall be protected so far as is reasonably practical against dangers that may arise from contact with live parts of the installation.

Protection under normal conditions, or in other words protection against electric shock when there is no fault, is called *basic protection*. This type of protection will prevent electric shock by enclosing all live parts or insulating all live parts.

Protection against electric shock under fault conditions is called *fault protection*. Automatic disconnection of the supply in a determined time in the occurrence of a fault will provide fault protection. This of course will usually include protective earthing and protective equipotential bonding.

Other methods used can be double or reinforced insulation, electrical separation and extra low voltage, all as described in Chapter 41 of BS 7671.

Practical Guide to Inspection, Testing and Certification. 978-1-138-61332-4

The most common method of fault protection used within a normal electrical installation is by the use of protective equipotential bonding and the automatic disconnection of the supply (ADS). Class 2 equipment (double or reinforced insulation), SELV and PELV and electrical separation (shaver socket) are also very common methods.

In a singe phase system current flow is achieved by creating a difference in potential.

If we were to fill a tank with water and raise the tank a metre or so, then connect a pipe with a tap on one end of it to the tank, when we open the tap the water will flow from the tank to the open end of the pipe. This is because there is no pressure outside of the pipe; the higher we raise the tank the greater the pressure of water and therefore the greater the flow of water.

Current flow is very similar to this. If we think of voltage as pressure then to get current to flow we have to find a way of creating a difference in pressure. This pressure in an electrical circuit is called potential difference and it is achieved in a single phase system by pegging the star point of the supply transformer to earth. The potential of earth is known to be at 0 volts.

If we place a load between a known voltage and earth, the current will flow from the higher voltage through the load to earth. If we increase the voltage then more current will flow, just as more water would flow if we increased the height of the water tank.

The problem we have with electricity is that if we use our body to provide the current with a path to earth, it will use it, and possibly electrocute us at the same time.

Current will not flow unless it has somewhere to flow to, and that is from a high pressure to a lower pressure, possibly zero volts but not always. It is also possible in some instances to get different voltages in an installation, particularly during a fault where volt drops may occur due to loose connections, high resistance joints and different sizes of conductors. We must also remember that during a fault it will not only be the conductors that are live but any metalwork connected to the earthing and bonding system, either directly or indirectly. It is highly likely that an electric shock could be received between pipe work at different voltages.

In any installation protection must be in place to prevent electric shock. The protection we use against someone touching a part which is intended to be live is self-explanatory, we can only prevent unintentional touching of live parts. Where a person is intent on touching a live conductor we can only make it difficult for them, not impossible.

Protection against electric shock is a different problem altogether and we can achieve it by different methods.

First, if there is a fault to earth all of the metal work connected to the earthing system, whether directly or indirectly, would become live. In the first instance we need to ensure that enough current will flow through the protective device to earth to operate the protective device very quickly. This is achieved by selecting the correct type of protective device, and ensuring that the earth fault loop path has a low enough impedance to allow enough current to flow to operate the device in the required time. On its own this is not enough and that is where the equipotential and supplementary bonding is used. The basic principle is that if one piece of metal work becomes live, any other parts that could introduce a potential (voltage) difference also become live at the same potential. If everything within the building is at the same potential current cannot possibly flow from one part to another via a person or livestock.

Ingress protection

In BS 7671 Wiring Regulations the definition of an enclosure is 'a part providing protection of equipment against certain external influences and in any direction providing basic protection'.

To ensure that we use the correct protection to suit the environment in which the enclosure is installed, codes are used. These codes are called IP codes. IP is for ingress protection and is an international classification system for the sealing of electrical enclosures or equipment.

The system uses the letters IP followed by two or three digits; the first digit indicates the degree of protection required for the intrusion of foreign bodies such as dust, tools and fingers.

The second digit provides an indication of the degree of protection required against the ingress of moisture.

If a third digit is used, a letter will indicate the level of protection against access to hazardous parts by persons; a number is used to indicate the level of protection against impact.

Where an X is used it is to show that nothing is specified; for example, if a piece of equipment is rated at IPX8 it would require protection to allow it to be submersed in water. Clearly if a piece of equipment can be submersed safely then dust will not be able to get in to it and no protection against the ingress of dust would be required.

The third number for impact is not used in BS 7671 and is not included in this book.

Table 9.1 Table of IP ratings

Dust and foreign bodies	Level of protection	Moisture	Level of protection
0	No special protection	0	No special protection
1	50mm	1	Dripping water
2	12.5mm diameter and 80mm long (finger)	2	Dripping water when tilted at 15°
3	2.5mm	3	Rain proof
4	1mm	4	Splash proof
5	Limited dust	5	Sprayed from any angle (jet proof)
6	Dust tight	6	Heavy seas and powerful jets
		7	Immersion up to 1M
		8	Submersion 1M +

Table 9.2 Third letter

A	The back of a hand or 50mm sphere
B	Standard finger 80mm long
C	Tool 2.5mm diameter, 100mm long, must not contact hazardous areas
B	Wire 1mm diameter, 100mm long, must not contact hazardous areas

CHAPTER 10

Testing photovoltaic systems

Photovoltaic systems are becoming more and more common and it is important that any electrician can inspect them and test them. It is not a specialist area of work and is certainly within the skills set of a competent electrician.

The a.c. side of these installations will require an electrical installation certificate to be completed along with the required schedules; this is no different from the requirements for any new circuit.

The d.c. side of the installation must also be inspected with some tests being carried out to verify that it is operating correctly.

Testing and commissioning

As with any other electrical installation it is a requirement that the installation is inspected, tested and commissioned by a competent person and that the correct certificates are completed and handed to the customer along with the user instructions provided by manufacturers.

As with any commissioning a visual inspection is important and is the first part of the process.

Visual inspection

Panels

- Are the PV panels are to British Standard?
- Are the PV panels correctly fixed to the roof or other part of the building?
- Where the panels form part of the roof, is the roof weathered properly with the correct type of flashings installed where required?

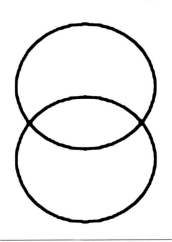

Figure 10.1 Transformer fitted

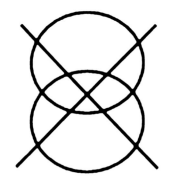

Figure 10.2 Inverter without transformer

- Where the panels are retrofit and are fitted on the roof, is there suitable ventilation beneath the panels?
- Is the roof suitable for the additional load? In most cases it is better to have a roof survey carried out, as it is not only the weight of the panels which has to be considered. Wind load is also a major consideration. There are organisations which will carry out desk top appraisals for roof structure. The following link may be of some use: www.structuralreportsonline.co.uk/structural-calculations-for-solar-panel-installation.
- Does the inverter have simple separation between the d.c. and a.c. side? This can be found by reading the manufacturer's instructions. Some inverters will have a symbol on the side which will indicate whether the inverter has an isolation transformer or not (Figures 10.1 and 10.2).

Any system which has an inverter which does not provide simple separation must have the array frame bonded; the bonding can be to the main earth terminal for TT and TNS systems. If the supply system is a TNCS system the array frame must be bonded using an earth electrode.

Cables

- Are the cables from the panels tied to the array frame? Remember that they will be expected to be in place for 25 years. Roof tiles can be very abrasive and if the cables are rubbing over them on windy days, the insulation will soon wear away.
- Where the cables penetrate the roof, the cables are not bent with too tight a radius, and there are no sharp edges which could damage the cables.
- Are the cables in the roof properly supported (clipped or in a PVC enclosure)?
- Are the correct type of cable connectors used?
- Are the correct type of cables used for d.c. and a.c. installation?
- Is the cable correctly selected for current rating and voltage drop?
- Has mechanical protection been provided for cables?
- Are the d.c. cables correctly labelled for identification purposes?
- Have string fuses been fitted where four or more strings have been installed?

Control equipment

- Has a d.c. isolator been provided before the inverter on the d.c. part of the installation? This device need not be lockable, although it is preferable but in the off position only.

- Is the inverter correctly rated and manufactured to an appropriate standard?
- Has an a.c. isolator been fitted on the a.c. side of the inverter? This device must be lockable in the off position only.
- Is all equipment correctly labelled?
- If the installation is going to be connected to the grid and the feed in tariff claimed, are all products used compliant to MCS standards (Microgeneration certification scheme)?
- Is there an a.c. isolator fitted next to the consumer unit to provide complete isolation of the PV system from the electrical installation?

General

Labelling and identification is required throughout the installation to reduce the risk of accidents being caused due to misidentification. Labels would be required for the following:

All conductors

- On the d.c. side of the installation, are the conductors correctly identified (Figure 10.3)? The positive must be brown and the negative is grey. A small length of coloured sleeving or coloured tape is suitable.
- On the a.c. side of the installation, positive is brown and neutral is blue.

Isolation points

- d.c. side of inverter
- a.c. side of inverter
- warning of dual supply, indicating isolation point for PV and mains supply (Figure 10.4)
- main isolation a.c. isolation point.

Live DC Cable
Do not disconnect DC plugs under load.
Turn off AC and DC isolators first.

Figure 10.3 D.C. identification

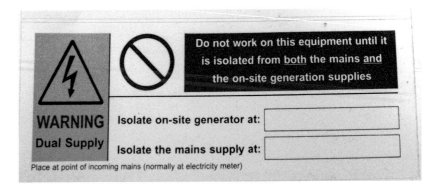

Figure 10.4 Dual isolation label

Warning signs

- PV junction boxes (live in daylight).
- Do not disconnect d.c. plugs and sockets under load.

A check must also be made to ensure that all of the equipment is rated at a minimum of 1.15 times the maximum voltage and 1.25 times the maximum current which is produced in the system.

Commissioning test sheets must be completed showing which items have been inspected and which items on the sheet are not applicable to the particular installation (Figure 10.5).

On completion of the visual inspection the installation must be tested to ensure that it is safe and suitable for use. When carrying out testing on PV installations it is very important to remember that the d.c. side of the installation will be continually live and will present a risk of electric shock.

Documentation must be provided to show that both the d.c. and the a.c. parts of the microgeneration system have been satisfactorily tested.

Testing the d.c. side of the installation

The open circuit voltage (V_{oc}) of the installation must be measured and recorded. This is simply a matter of measuring the voltage across the positive and negative cables at the d.c. isolator nearest to the inverter (Figure 10.6). This is really just a check to make sure that the system is working to its full potential.

For example if we had 6 panels each with a V_{oc} of 24 volts we would expect a terminal voltage at the inverter of 144 volts. This would prove that the array was working correctly.

Megger.

Certificate No: 2

PV System Inspection Report

Client	Contractor
Address	Address

BS7671 inspection report reference	PV array test report reference	Intitial verification
BS7671 test report reference	Rated power - kW DC	Periodic verification
PV array inspection report reference	Location	Date
Description of installation	Circuits tested	

General

☐ The entire system has been inspected to the requirements of IEC 60364-6 and an inspection report to meet the requirements of IEC 60364-6 is attached.

PV array design and installation

☐ DC system designed, specified and installed to the requirements of IEC 60364 in general and IEC 60364-7-712 in particular.

☐ DC components rated for continuous DC operation.

☐ DC components rated for current and voltage maxima (Voc stc corrected for local temperature range and module type; current at Isc stc x 1,25 - IEC 60364-7-12.433:2002).

☐ Protection by use of class II or equivalent insulation adopted on the DC side   (class II preferred - IEC 60364-7-712.413.2:2002).

☐ PV string cables, PV array cables and PV DC main cables have been selected and erected so as to minimize the risk of earth faults and short-circuits (IEC 60364-7-712.522.8.1:2002).

☐ Wiring systems have been selected and erected to withstand the expected external influences such as wind, ice formation, temperature and solar radiation (IEC 60364-7-712.522.8.3:2002).

☐ Systems without string over-current protective devices: String cables sized to accommodate the maximum combined fault current from parallel strings (IEC 60364-7-712.433:2002).

☐ Systems with string over-current protective devices: over-current protective devices are correctly specified to local codes or to the PV module manufacturers instruction - to NOTE of IEC 60364-7-712.433.2:2002).

☐ DC switch disconnector fitted to the DC side of the inverter (IEC 60364-7-712.536.2.2.5:2002).

☐ If blocking diodes are fitted, verify that their reverse voltage rating is at least 2 x Voc stc of the PV string in which they are fitted (IEC 60364-7-712.512.1.1:2002).

☐ If one of the DC conductors is connected to earth, verify that there is at least simple separation between the AC and DC sides and that earth connections have been constructed so as to avoid corrosion (IEC 60364-7-712.312.2:2002).

PV system - protection against overvoltage / electric shock

☐ If an RCD is installed and the PV inverter is without at least simple separation between the AC side and the DC side: is the RCD of type B according to IEC 60755 (IEC 60364-7-712.413.1.1.1.2:2002 and Figure 712.1).

☐ Area of all wiring loops has been kept as small as possible (IEC 60364-7-712.444.4:2002).

☐ Array frame equipotential bonding has been installed (to local codes)

☐ Where installed, equipotential bonding conductors are laid parallel to and bundled with the DC cables.

PV system - AC circuit special considerations

☐ Means of isolating the inverter have been provided on the AC side.

☐ Isolation and switching devices have been connected such that PV installation is wired to the "load" side and the public supply to the "source" side (IEC 60364-7-712.536.2.2.1:2002).

☐ Inverter protection settings are programmed to local regulations.

PV system - labelling and identification

☐ All circuits, protective devices, switches and terminals are suitably labelled.

☐ All DC junction boxes (PV generator and PV array boxes) carry a warning label indicating that active parts inside the boxes are fed from a PV array and may still be live after isolation from the PV inverter and public supply.

☐ Main AC isolator is clearly labelled.

☐ Dual supply warning labels are fitted at point of interconnection.

☐ Single line wiring diagram is displayed on site.

☐ Inverter protection settings and installer details are displayed on site.

☐ Emergency shutdown procedures are displayed on site.

☐ All signs and labels are suitably affixed and durable.

PV system - general installation (mechanical)

☐ Ventilation provided behind array to prevent overheating / fire risk.

☐ Array frame material corrosion proof.

☐ Array frame correctly fixed and stable; roof fixings weatherproof.

☐ Cable entry weatherproof.

DESIGN, CONSTRUCTION, INSPECTION AND TESTING

I/we being the person(s) responsible for the design, construction, inspection and testing of the electrcial installation (as indicated by the signatures(s) below), particulars of which are described above, having exercised reasonable skill and care when carrying out the design construction, inspection and testing, hereby certify that the said work for which I/we have been responsible is, to the best of my/our knowledge and belief, in accordance with

Date   Next inspection recommended after not more than

Comments

Signature

Name

(The extent of liability of the signatory(s) is limited to the work described above)

Figure 10.5a Commissioning report

Certificate No: 2

Megger.

PV Array Test Report

Installation Address								Initial verification

Periodic verification

Postcode

Reference

Description of work under test

Date

Inspector

Test Instruments

String						
Array	Module					
	Quantity					
Array Parameters	Voc (stc)					
	Isc (stc)					
String overcurrent protective device (1)	Type					
	Rating (A)					
	DC rating (V)					
	Capacity (kA)					
Wiring (1)	Type					
	Phase (mm^2)					
	Earth (mm^2)					
String tests	Cell Temperature (2)					
	Irradiance					
Calculated values (3)	Voc expected (V) (3)					
	Isc expected (A) (3)					
Measured values	Voc actual (V)					
	Isc actual (A)					
Array insulation resistance	Test voltage (V)					
	Pos - Earth (MΩ)					
	Neg - Earth (MΩ)					
Earth continuity (where fitted)						
Switchgear functioning correctly						
Inverter make / model						
Inverter serial number						
Inverter functions correctly						
Loss of mains test						
Generation meter serial number						
Generation meter kWh reading						

Comments

Notes

(1) These values are recorded in the schedule of test results, for the purpose of City & Guilds 2399 they do not also need to be recorded here in the PV array test report.

(2) The cell/module temperature needs to be measured in order to calculate the expected Voc and has been added to the City & Guilds 2399 version of this form.

(3) The expected Voc & Isc need to be calculated in order to validate the measure values, and have been added to the City & Guilds 2399 version of this form.

Page 2 of 2

Figure 10.5b Continued.

Unfortunately it is not quite as simple as it seems because the maximum V$_{oc}$ rating given for each PV panel will be the voltage which it would produce under standard test conditions (STC). This involves the manufacturer subjecting the panel to a set level of irradiance at a known temperature. The standard test conditions are an irradiance 1000w at a temperature of 25°C with an air mass of 1.5. Of course once the panels have been installed, the levels of irradiance that the panels will be subject to and the temperature of the panels will change constantly.

Experience is the easy way to check if the array voltage is as expected. On a really bright day the voltage will be close to that which has been calculated, on an overcast day the voltage will be slightly less.

As an example we can use values from Table 10.1, which is typical of the information provided on a data sheet for a photovoltaic panel.

Table 10.1 Panel data sheet

Rated maximum power	180W
Tolerance	−0/+3%
Voltage at Pmax (Vmp)	36.4V
Current at Imp (Imp)	4.95A
Open-circuit voltage (V$_{oc}$)	44.2V
Short-circuit current (I$_{sc}$)	5.13A
Nominal operating temp	45°C
Maximum system voltage	1000V d.c.
Maximum series fuse rating	15A
Operating temperature	−40°C to +85°C
Protection class	Module application Class A
Cell technology type	Mono – SI
Weight (kg)	15.5
Dimensions (mm)	1580 X 808 X 45
Standard test conditions	Cell temp = 25°C AM 1.5 1000W/m²
Temperature coefficient (V$_{oc}$)	−0.172 V/°C
Temperature coefficient (I$_{sc}$)	0.88 mA/°C

Figure 10.6 PV array test report

Voltage test on a photovoltaic array

Video footage is also available on the companion website for this book.

We can see that the V_{oc} (max voltage) is 44.2 volts per panel. This means that if we have 6 panels connected in series the total voltage would be $6 \times 44.2 = 265.2$ *volts*.

This of course is the maximum voltage which will only be present on a perfect day. A change in irradiance does not have a great effect on the voltage produced by a panel, and during daylight hours the voltage will remain reasonably constant although on very dark gloomy days the effect will be greater.

Apart from very dark days the greatest effect on the voltage produced by the panels is due to temperature. Table 10.1 shows that there is information on the temperature coefficient (V_{oc}) of the panels. The value given is 0.172v/°C. This means that the voltage will alter by 0.172 volts for each degree change in temperature.

When completing the commissioning sheet for the d.c. side of a PV system the voltage must be tested to ensure that it is correct. The temperature must be taken into account during this test.

Example 1

We must collate some information before we start. This can be read from the data sheet in Table 10.1.

V_{oc}. *44.2 The maximum amount of voltage which could be produced by the panel.*

STC. 25°C The temperature at which the panel was originally tested.

Temp Coefficient. 0.172v/°C The change in voltage per deg.

As well as this information we also need to measure the temperature of the panel. This can be done by using an infrared temperature sensor, or in some installations the inverter will provide this information. For our example let's say that the temperature of the panel is 35°C.

As with any calculation there are various ways which it can be carried out.

To explain it before we start, the voltage output of the panel will rise or fall due to a change in temperature. The higher the temperature the lower the voltage; in other words high temperatures will have a negative effect on the output of PV panels. In our example the voltage will reduce by 0.172 volts for each degree centigrade rise in temperature.

The standard test temperature for the panels is 25°C and our actual panel temperature is 35°C; this of course is a rise in 10°C.

If the panel voltage reduces by 0.172 volts per degree then we need to multiply this value by 10.

$$10 \times 0.172 = 1.72v$$

This tells us that the voltage output of the panel will now be

$$44.2 - 1.72 = 42.82 \text{ volts}$$

Which of course results in the total voltage produced reducing from

$$6 \times 44.2 = 265.2 \text{ } volts \text{ to } 6 \times 42.82 = 256.92 \text{ } volts$$

This of course will have an effect on the power output of the system, however it is worth remembering that on cold days the voltage will increase by the same proportions.

A far greater influence on the output of the panels is the irradiance (amount of sunlight). Table 10.1 shows us that the standard test conditions for the panel were at $1000wm^2$.

The amount of energy produced by the sunlight on a good day is considered to be 1kW per m^2. This of course will vary slightly in different parts of the country; in the south west this can rise to $1.2kW^2$.

The panels are all tested at the same irradiance as it is important for us to be able to make a fair comparison between panels; $1000wm^2$ is the STC for all panels.

From Table 10.1 we can see that the max short circuit current (I_{sc}) is 5.13A – this of course is at an irradiance of $1000wm^2$.

To prove that the system is working correctly we must compare the short-circuit current to the amount of the irradiance. To do this we must measure the irradiance levels using an irradiance meter (Figure 10.7) and at the same time measure the short-circuit current (I_{sc}) of the array using a d.c. clamp meter (Figure 10.8).

On a bright sunny day the measurements can be taken separately as the irradiance levels will be reasonably constant over a short period of time. However on a cloudy day getting an accurate measurement can prove difficult, and may involve two people. The irradiance value can change instantly and this of course will have an effect on the current being produced.

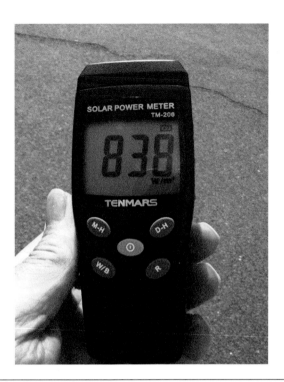

Figure 10.7 Irradiance meter

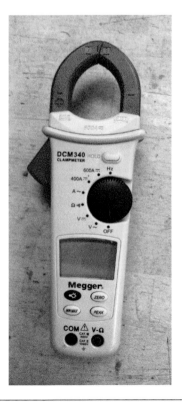

Figure 10.8 D.C. clamp meter

Example 2

Let's say that the measured irradiance is 80 wm².

Table 10.1 shows us that the short circuit current Isc is 5.13A (at 1000wm²).

Having got both measurements the calculation is as follows:

Measured irradiance value = 800 W/m² now divide this by 1000 to convert to kW = 0.80 kW

The standard test current (Isc) for panel is: 5.13A (if the panels are in a string the current will not increase).

Multiply the total stc by the measured irradiance in kW

$$5.13 \times 0.80 = 4.14A$$

The measured short circuit current should be approximately 4.14A.

There are two ways to measure the short circuit current. It is a simple process but REMEMBER the d.c. side of the installation will be live during daylight hours.

Method 1

Isolate the complete pv installation.

Isolate the d.c. side of the installation by using the d.c. isolator nearest to the inverter.

Insert a shorting link across the positive and negative terminals on the dead side of the isolator, or join male and female connectors (Figure 10.9).

Switch on the d.c. isolator and place a d.c. clamp meter around one of the incoming d.c. cables and record the reading (Figure 10.10).

Switch off the d.c. isolator and remove links.

Switch the installation back on and leave it in the operating condition.

Method 2

This method requires the use of a d.c. current meter; most good quality multimeters will be suitable. Set your multimeter at a current which is higher than the current expected, place test probes across the positive and negative terminals and take a reading. If the reading is far lower than the setting of your instrument it is better to switch the instrument to a lower setting to get a more accurate measurement (Figure 10.11).

Figure 10.9 D.C. leads joined

Protective earthing of the PV installation is normally not a requirement as the equipment used is usually class 2 equipment; however, to reduce the risk of electric shock an insulation resistance test should be carried out between the live conductors and live conductors to earth.

Insulation resistance test on a PV array

Video footage is also available on the companion website for this book.

Insulation resistance test

For this test an insulation resistance test instrument is used set at a voltage of 500 v d.c. on a new installation. This test is to be carried out between live conductors and earth, and between live conductors before the panels are connected.

On an existing installation the test between live conductors is not carried out as the conductors will be live during daylight hours.

Two methods can be used for testing existing installations: one method is for the positive and negative of the pv installation to be connected together and a test must be carried out between these

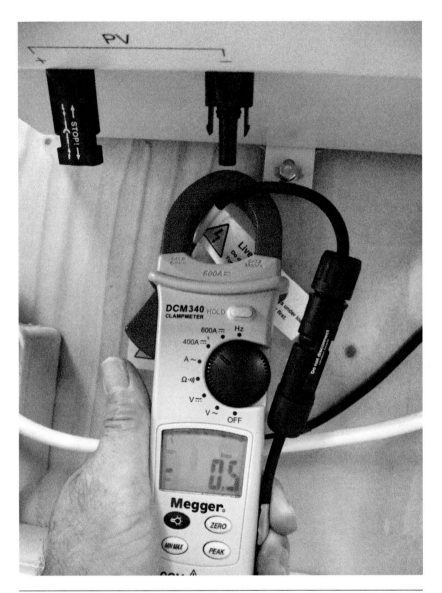

Figure 10.10 Current measurement using the clamp meter

and any metal parts of the installation. The d.c. conductors will be live during this test, for that reason it is important that the d.c. side is isolated.

Method 1

Step 1

Isolate d.c. side (Figure 10.12).

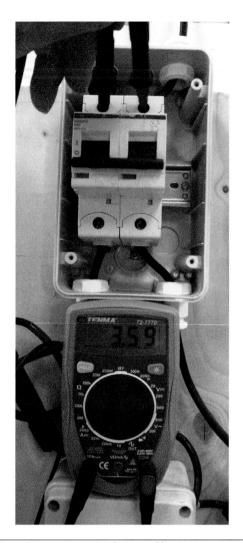

Figure 10.11 Current measurement using a multimeter

Step 2

After disconnecting out going cables link the two d.c. terminals on dead side of isolator (Figure 10.13).

Step 3

Connect insulation resistance leads to dead side of isolator and a known earth (Figure 10.14)

Step 4

Switch on isolator and measure insulation resistance

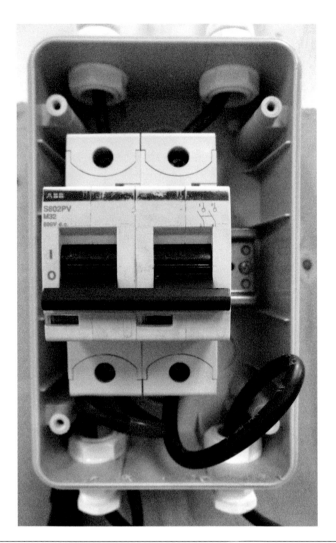

Figure 10.12 D.C. isolated

Step 5

Isolate and remove leads, reconnect cables.

A second test must be carried out between the joined conductors and the main installation earth.

Method 2

The other method is to carry out the test using the positive and negative separately.

Where the test is being carried out as described in the second method precautions must be taken to ensure that the test voltages do not exceed the module or cable rating.

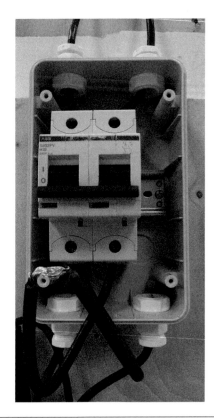

Figure 10.13 D.C. cables linked

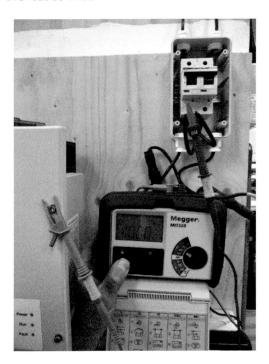

Figure 10.14 Insulation resistance test

Testing the a.c. side of the installation

For all new connections to a supply system, an electrical installation certificate must be carried on the a.c. part of the PV installation. On completion an Electrical Installation Certificate (see Figures 2.1 and 2.2) must be completed, along with a schedule of test results (see Figure 6.3) and a schedule of inspections (see Figures 2.3 and 2.4). These documents must only be completed by a person who is registered to self-certificate electrical work which has been carried out by them.

In installations where it has been chosen to connect the PV circuit by using a spare way in an existing consumer unit, it is a requirement that the earthing arrangements for the existing installation are upgraded to comply with the latest edition of BS 7671 Wiring Regulations.

Fault finding

As far as I am concerned the most effective method of fault finding is to work methodically – this, combined with experience and a bit of luck, usually ends up with the correct result. Of course not all faults are the same and some are very difficult to find, but once found they are usually quite simple to correct.

I find that it is always better to step back and think about what you are doing, and not jump in with both feet and start taking things apart. The very first step is to gain as much evidence/information as possible, ask the customer or the person who has discovered a fault to provide as much information as they can as this will help.

I always ask if anyone else has been trying to find the fault, has anything been disconnected? This information could have a massive effect on the way you approach a fault; for instance if something has been working, but will not work now and has not been touched other than to switch it on and off, then it is a genuine fault and something in the circuit has failed. If however someone has been pulling it around and disconnecting things that they know nothing about, you will have a different job on your hands as you cannot be sure that it is connected correctly.

Ring final circuit

Take the example of a ring – when we measure end to end values there is no continuity between the ends of the line conductor but all of the other conductors measure as expected.

Of course the break could be at any point in the ring and is probably just a cable snapped at one of the terminations. As with most fault finding we need a bit of luck mixed with our knowledge; the most important thing to remember is to work methodically.

It could be that we have not installed the ring and have no idea as to the route of the cables; we now have to just take a calculated guess as to how the ring is wired.

Take a socket off somewhere around the middle of the ring.

- At the board link L and E of one end of the ring.
- At the socket test between L-E of one of the ring ends using a low resistance ohmmeter.
- If you have continuity go to the board and split the joined ends, but mark which ones they were.
- Now test L-E again at socket and if the circuit is open resistance you will know that the break is not on the end of the ring which you have tested. Again mark the end which is ok.
- Now join the L-E of the other end of ring at the board.
- Test L-N at the socket of the unmarked end and the circuit should be open circuit.
- If it is open circuit join the L-E of the end just tested.

Now comes the tricky bit as you have to guess which part of the ring has the break. You need to take a chance on which way the ring has been run, and take a socket off somewhere between the board and the socket.

- Disconnect conductors from the socket.
- Test between L-E of both cables.
- If both ends are open circuit there is a good chance that you are working on the wrong side of the ring. The best way to find out is to join L-E of the ends which you marked at the socket and board, test again at the socket and you should have a closed circuit.
  If that is the case then just put the socket back on and choose another somewhere else on the ring.
- If however when you take the socket off and measure the ends one of them is closed circuit and the other is open circuit you will be looking in the right place.
- Now you need to guess which way the cable which is open circuit runs; it may go to the board or it may go to the first socket which you removed.
- Take another socket off between the socket you have just tested, and either the first socket or the board. This is your choice and also a bit of luck is needed.
- You need to keep doing this until the fault has been found; it is just a process which keeps halving the faulty section.

Of course once the fault has been found then all of the connections need to be remade and a complete ring final circuit test will need to be carried out.

Interpretation of ring final test results

Let's assume that we have a ring circuit wired in 2.5mm^2 thermoplastic 70°C twin and earth cable; the CPC is going to be 1.5mm^2.

The test between the ends of the line conductor gives a reading of 0.6Ω, from this value we will know to expect a value of 0.6Ω when we measure the N conductor end to end and a value of 1.67 times 0.6Ω ($0.6 \times 1.67 = 1\Omega$) when we measure the CPC end to end. If the values differ from these by very much then immediately we should recognise that there is probably a loose connection (high resistance joint) in one of the sockets; before we proceed with the rest of the test we should correct the problem.

Once corrected we will have values of R_1 0.6Ω. R_n 0.6Ω and R_2 1Ω.

When we cross-connect the line and N conductors we know from these values that we will expect a resistance value between L-N at each socket of $\dfrac{0.6 + 0.6}{4} = 0.3\Omega$. In reality though we will know that the expected resistance value will be half of the line conductor resistance.

When we cross-connect the line and CPCs we will expect a value of $\dfrac{0.6 + 1}{4} = 0.4\Omega$. Always remember though that as the CPC is a smaller CSA than the line conductor the value will be slightly lower on the sockets nearer to the cross-connection but will gradually increase to 0.4Ω as you measure the sockets nearer to the centre of the ring.

Table 11.1 gives some measured values which if we interpret correctly will give us a good idea of how a problem could be identified.

Now let's look at how these values have been interpreted to find the possible faults/inconsistencies.

Socket 1. We can see that the values are as expected.

Socket 2. As we look at a socket face (A) in Figure 11.1 we expect the connections to be as shown.

If the connections were as shown in (B) the reading when measured between L-E would be as expected but would be open circuit between L-N proving that the socket has been connected incorrectly or that the N conductors have dropped out of the terminal.

Socket 3. The value shows us that it must be a loose N connection as L-E are as expected.

Socket 4. As the values are both high it will show either a spur or a loose connection on the conductor which is common to both tests; this of course is the line conductor.

Table 11.1 Measured values

L-L = 0.6Ω	N-N = 0.6Ω	CPC-CPC = 1Ω	
L – N cross connected 0.3Ω		L – CPC cross connected 0.4Ω	
Socket	L-N	L-E	
1	0.3Ω	0.4Ω	Correct
2	No reading	0.4Ω	Reverse L-E
3	0.6Ω	0.4Ω	High resistance N connection
4	0.6Ω	0.8Ω	Possible spur or high resistance L connection
5	0.3Ω	0.7Ω	High resistance E connection
6	0.3Ω	No reading	Reverse L – N
7	No reading	No reading	Reverse N – E

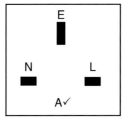 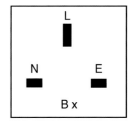

Figure 11.1 Socket 2 A and B

Socket 5. Ok between L-N but high between L-E so must be loose earth connection.

Socket 6. No reading between L-E but a correct reading between L-N which will show that L – N are reversed as shown in socket (B) (Figure 11.2).

Socket 7. No reading at either measurement must show that the N-E are reversed as shown (Figure 11.3).

All of these mistakes are very easy to make and it is a good reason to carry out all tests correctly as human error is very difficult to eliminate.

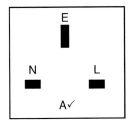 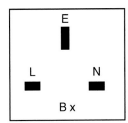

Figure 11.2 Socket 6 A and B

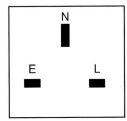

Figure 11.3 Socket 7

RCD tripping

If the circuit is protected by an RCBO, then of course the first step is to disconnect as much of the equipment connected to the circuit as possible. If it is a power circuit then unplug everything that you can find, switch off fused connection units then try to reset the RCBO. If it will not reset, isolate the circuit and disconnect the line conductors but not the CPC. The CPC should be left connected to the earth bar as the fault may be between the circuit live conductors and another exposed or extraneous conductive part, not necessarily between the live conductors and the CPC.

Now carry out an insulation resistance test between the live conductors joined together and the earth bar; I usually test at 250v first just in case there is surge protection. Now split the circuit around the centre, this will indicate which side of the circuit has the fault. Split the faulty side of the circuit again until you have found where the fault is. Of course it is not always as easy as that as the fault may be under the floor (crushed cable, rodent damage) but at least the test will narrow down the area for you.

If the fault is on a lighting circuit, switch off all of the lights and then try to reset the RCBO. If it resets, then just switch on the lights one at a time until you find the one which trips the RCBO.

If however the RCBO will not reset then have a good look around for any switches which look like they may have been disturbed, or light fittings which have been recently fitted. If there is nothing obvious then isolate the circuit and disconnect the live conductors.

Where installations are protected by an RCD main switch, the obvious process is to switch off all of the circuit breakers or remove all of the fuses. Switch on the RCD and replace the protective devices one at a time until the RCD trips, remove the device which trips the circuit and replace all of the others just to be sure that the fault is only on one circuit. This will then identify the circuit and then you can begin the fault finding.

If the RCD will not reset when all of the protective devices have been removed, it will indicate that the fault is a neutral to earth fault. The procedure now is to remove the neutrals one at a time and carry out an insulation resistance test between the disconnected N and earth until the faulty circuit is found. Once found, replace all of the other disconnected neutrals and protective devices and switch on the RCD; if it stays on then you have found the faulty circuit.

Something to be aware of when fault finding and RCDs are involved is a situation that often arises when RCBOs are used along with an RCD main switch. Occasionally the main RCD will trip, as well as the RCBO even if the main switch is time delayed, or higher rated.

This is because most RCBOs are single pole and although they identify a N to E fault and disconnect, the N remains connected and in turn causes the RCD to trip.

When upgrading consumer units to comply with BS 7671 and a split board is used with 2 or 3 RCDs, it is important to ensure that the neutrals and line conductors are connected to the same section of the board. If a mistake is made you will soon discover that when something is connected to the circuit the RCD will trip; this of course will look like a faulty circuit.

Insulation resistance

When insulation resistance testing, always take care as the test is usually carried out at 500 volts d.c. The test leads should comply with GS38 and it is really important that the tester is checked for correct operation by connecting the leads together and pushing the test button. This should provide a reading of 0.00. Now part the leads and repeat the test and a reading of the maximum resistance should be seen such as >300MΩ but this will depend on the tester being used.

Remember that this is a dead test and the circuit or complete installation must be isolated before testing can begin.

Let's assume that we are going to test a complete installation for a condition report and that isolation has been achieved by switching off at the main switch.

Now we must look around the building for any fixed appliances which may be switched on; any found should be switched off.

Unplug anything which is plugged into a socket outlet. Isolate any controls for the heating/hot water system; this is usually just a matter of switching off or unplugging the heating control supply.

Check that any neon indicator switches are turned off as well as any equipment which may be damaged by the test voltage, this could include motion sensors and dimmer switches. These should be either bypassed or linked across.

Remove any lamps where possible. If discharge lamps are used such as fluorescent fittings which have control gear fitted, or even any lamps which are difficult to access, just turn off the switch which would operate them.

What we have done is to try to remove any equipment which may result in giving us a low reading due to our test current passing through it.

With the main switch off and all of the circuit breakers on or protective devices in place we can begin the test.

The first test which I would carry out is between live conductors and I would always start at 250 volts; this is because if any equipment which we have missed but which could be damaged at 500 volts will show up as a bad reading but will not be damaged.

If the test results in a low reading then I would look around again just to check for anything which has been missed; a simple method to narrow the search would be to turn off all of the circuit breakers or remove protective devices. Carry out the test again but switch on the circuit breakers/replace devices one at a time; this should identify the circuit with the problem.

Identifying a lost switch line on a three plate lighting circuit

I am sure that I am not the only person to get numerous calls throughout the year after someone has taken down a light fitting, and find that when they reconnect it there is a bang! The story is usually that they have connected all of the blues/blacks together, and all of the brown/reds together and when the light is switched on all of the other lights go out.

I find the simplest method is this.

Isolate the circuit (*it probably already is but by accident*). Don't ever take anyone else's word that a circuit is dead, always carry out the procedure yourself and always lock off correctly. Better safe than sorry!

Once isolated, disconnect all of the blues/blacks, turn the switch on and connect a buzzer or low resistance ohmmeter to the browns/reds and one blue/black. Test each blue/black to the brown/reds until you get a reading. Once you have a reading you must now turn off the switch. If the reading disappears you have found the switch return; if it does not disappear you will most likely have found a live pair coming from another light. In this case move on to another blue/black.

Immersion heater not working

The faults on immersion heaters are usually quite easy to identify although there are many symptoms and various ways of identifying the faults.

Where the circuit is protected by an RCD the breakdown of an immersion heater element will usually trip it. Of course the easiest way is just to turn off the local isolator to the immersion heater and then turn on the RCD. If it stays on then it will show that the immersion heater is the problem.

Always be aware though that the isolator needs to be double pole; if it is not then the RCD may still trip as it will detect the leakage between earth and neutral.

Where an RCD is not part of the circuit a damaged element will not usually operate a circuit breaker, and the only symptom may be that the immersion heater is not working. This type of fault will require a methodical approach.

Always check the obvious – is the circuit switched on? If it is connected by a switched fuse connection unit (fused spur) is the fuse in the connection unit the correct size? Is it in good condition? If yes then switch off local isolation to the heater, remove the cover of the immersion heater and using a voltage indicator check that it is isolated.

Visually inspect the connections in the immersion heater. If it has an integral overheat trip, press the trip button. If it clicks it will mean that it has reset (see Figure 11.4).

Now turn on the local isolation and using a voltage indicator to GS38, test between the ends of the element within the immersion heater (Figure 11.5).

Figure 11.4 Reset overhead trip

Figure 11.5 Check for voltage

If the voltage indicator shows that the full voltage is present then it will indicate that the thermostat is closed and that the element has a supply to it.

If there is no supply to the heater element then test between the incoming supply to the thermostat and neutral (Figure 11.6). If there is a supply voltage then at least you will know that the circuit is working

Figure 11.6 Incoming supply and N

Figure 11.7 Clamp meter with immersion off

Figure 11.8 Clamp meter with immersion on

up to the heater. This will indicate that the heater thermostat may be faulty or perhaps has been turned down. Check the setting just in case – it should be at a maximum of 60°C.

Where there is a supply to the heater element, we will still not know whether or not it is heating up as it will take some time before there is a change in water temperature. It could be of course that the water has been heated by a boiler and has reached its maximum temperature. The best and easiest way is to check if it is working is to turn off the heater, and turn up the heater thermostat to its highest setting.

Place a clamp meter onto the incoming meter tail and take a note of the current being drawn (Figure 11.7). Now turn the immersion heater on and see if the current has increased by the rating of the heater, as an example a 3 kW heater will increase by 13amps (Figure 11.8).

If the immersion heater has gone open circuit and not damaged the casing of the element then it may not trip an RCD or operate a fuse or circuit breaker. The heater terminals will be live but will not show an increase of current when using the clamp meter. In this case the immersion heater will require a replacement to be fitted.

CHAPTER 12

Exercises and questions

Exercise 1

Mrs Volks of 3 Wagon Close, Bath, Somerset, SO3 6HT has had you install a new 45 ampere/230v shower circuit in the bathroom of her house.

The shower control unit is on the outside of the bathroom. The existing installation is in good condition and complies with BS 7671: 2008 amended 3:2015. A recent electrical condition report, schedules of inspection and test results are available.

The wiring for the new circuit is 10mm^2 thermoplastic (pvc) flat twin with CPC cable. The circuit protection is by a 50A type B circuit breaker to BS EN 60898 with an I_{pf} rating of 6kA. This protective device is housed in a split RCD protected consumer unit which is situated in the integral garage of the house. The main switch is a 30mA BS EN 61008-1 100A 230 V.

The ambient temp is 20°C; $r_1 + r_2$ value for this cable is 6.44 milli ohms per metre; the circuit is 32 metres long.

The supply is TN-S 230 V 50Hz, the main fuse is 100A BS 1361 and the measured values of PFC and Z_e are 800A and 0.24 ohms respectively. The supply tails are 25mm^2 copper, the earthing conductor is 16mm^2 copper and the main protective bonding to gas and water is 10mm^2 copper. Maximum demand is 85A.

The measured Z_s is 0.39Ω and an insulation test on all conductors shows >200 M ohms.

The RCD test shows operating times of 60ms @ × 1 and 17ms @ × 5 and the circuit is number 9 in the board.

A visual inspection shows no defects and the test instruments are:

- Low resistance ohm and insulation resistance meter Serial no. 08H46.
- Earth loop impedance tester Serial no. 076H90.

Using the information given, answer the following:

1 Complete a schedule of inspections.
2 Complete a schedule of test results.
3 Complete the appropriate certificate.
4 Using the rule of thumb, show if the measured value of Z_s is acceptable. Give a reason/ reasons why this measured value is lower than $Z_e + (R_1 + R_2)$.
5 If the shower unit had been a replacement unit, complete the minor works certificate.
6 Describe in detail how an earth loop impedance test would be carried out on this circuit.

Exercise 2

You have installed a new ring circuit consisting of seven socket outlets on the ground floor of a detached house for Mr Chewter of 2 Stonepound Road, Hurstpierpoint, West Sussex, BN8 Y33.

There is no documentation available for the existing installation but it appears to be in good condition; there was space available in the existing consumer unit for a new circuit.

The supply is a TNCS 230 volt 50 Hz single phase with an 100A supply fuse to BS 1361. Measured values of PSCC and Z_e are 1450A and 0.18Ω respectively.

On completion of the additional work the maximum demand is 88A.

Meter tails are 25mm² copper, the earthing conductor is 16 mm² copper and the main equipotential bonding conductors to the oil and water supplies are 10mm² (correctly connected).

The new circuit is wired in 53 metres of 4.0 mm²/1.5mm² thermoplastic (pvc) twin and earth cable. Protection is by a 30A BS 3036 semi enclosed rewirable fuse which has a maximum Z_s value of 1.04 Ω. The consumer unit, situated in a cupboard in the hall, has a 100A BS EN 61008-1 RCD as a main switch with an $I_{\Delta n}$ rating of 30 mA, a tripping time of 46 ms at its rating and 34 ms at 5 times its rating. The circuit is circuit number 7.

The insulation resistance value is >200MΩ.

A visual inspection of the new circuit shows no defects, and the test instrument used is a Megger multifunction instrument, serial number CJK 1047.

Using the information given, answer the following:

1 Complete the correct paperwork.

A few weeks after the circuit has been installed, Mr Chewter requests that you install an additional twin socket outlet in his office; this is to be spurred from the new ring circuit. The socket is 8 metres away from the nearest existing outlet.

The spur is to be $4.0mm^2/1.5mm^2$ twin and earth thermoplastic cable (pvc).

The resistance of $4.0 mm^2$ copper is 4.61 mΩ per metre and the resistance of $1.5mm^2$ copper is 12.10 mΩ per metre.

2 Complete the correct paperwork.

Exercise 3

Mrs Mary Mint of Polo House, Minto Close, Nutley, Kent, KT3 8JI is moving to a new address: 24 Trebor Road, Dover, Kent, KH10 1DR.

This is an existing bungalow which is 18 years old and requires the correct inspection and test to be carried out and relevant documents to be completed. Existing documentation is available; the last inspection and test was carried out 6 years ago along with some additions to the installation.

The circuits are shown in Table 12.1 (circuit breakers to BS EN 60898 with an I_{pf} of 6kA).

$R_1 + R_2$ per metre @ 20^c

$6mm^2$	3.08 m/Ωm
$4mm^2$	4.61m/Ωm
$2.5mm^2$	7.41m/Ωm
$1.5mm^2$	12.10 m/Ωm
$1.0mm^2$	18.10 m/Ωm

The supply is a 230v 50 Hz TN-S system with a Z_e of 0.63 ohms and a PFC of 1200A.

The main fuse is an 80A BS 1361 type 2. The consumer unit is under the stairs; the main switch is 100A 240v to BS 5773.

Table 12.1 Circuits details

Circuit	Protection	Live	CPC	Length	Max Z_s
1. Cooker	40A type B	$6mm^2$	$2.5mm^2$	23mts	1.09
2. Ring 1	32A type B	2.5	1.5	48	1.37
3. Ring 2	32A type B	4.0	1.5	53	1.37
4. I/H	16A type B	1.5	1.00	12	2.73
5. Lighting	6A type C	1.5	1.00	38	3.64
6. Lighting	6A type B	1.00	1.00	57	7.28

The earthing conductor is 6mm^2.

Protective bonding to gas and water is 10mm^2 and is correctly connected. Plumbing is all in copper tubing and there is no evidence of supplementary bonding.

All circuits have an installation value of >200 M ohms.

Circuit 3 has a PIR to control the outside lighting, and there is a shaver socket in the bathroom.

All circuits are installed using reference method 100.

The test instrument used is a Megger multifunction instrument, serial number CJK 1047.

Using the information given, answer the following:

1 Using rule of thumb for all circuits, show if the measured value of Z$_s$ is acceptable.
2 Complete the Electrical Installation Condition Report and test result schedule for this installation.

Exercise 4

Complete a schedule of test results for Table 12.2 with the circuit protective devices selected and fitted in the correct order for good working practise. Omit sections where no details are given but select the appropriate rating for the protective devices.

Installation method is 100.

Supply system is TN-S 230, volt measured Z$_s$ is 0.4Ω and PFC is 1.2kA.

All circuits are protected by BS EN 60898 type B circuit breakers with an I$_{cn}$ of 10kA.

Table 12.2 Circuit details

Circuit description	Phase conductor	CPC conductor	Circuit length
1. Lighting	1.5 mm^2	1 mm^2	32 metres
2. Ring	2.5 mm^2	1.5 mm^2	68 metres
3. Ring	4 mm^2	1.5 mm^2	72 metres
4. Shower	6 mm^2	2.5 mm^2	14 metres
5. Immersion	2.5 mm^2	1.5 mm^2	17 metres
6. Lighting	1 mm^2	1 mm^2	43 metres

Exercise 5

A remote cottage has been re-wired in flat twin and earth 70°C cable; all cables are concealed within the building fabric where possible. The supply system is a 230 volt single phase TT system and the main switch is a 30mA RCD. The earth fault loop impedance needs to be checked to confirm the suitability of the consumer earth electrode.

Using the information given, answer the following:

1 Which test instrument is to be used to carry out the test?
2 Which guidance document states the requirement for test leads and probes?
3 Describe how the test should be carried out.
4 What is the maximum earth fault loop impedance value permitted for this installation? Show all calculations.
5 What is the value of electrode resistance above which the value of the electrode resistance is considered unstable?
6 Which other method could be used to test the resistance of the earth electrode?
7 Which instrument is to be used for the test?
8 How many additional electrodes are required?

Exercise 6

An electrical installation within a new village centre forms part of a 400/230 v TN-C-S system. The measured values of PFC and Z_e have been recorded as 3.7kA and 0.12Ω.

The distribution boards and main switch are within a cupboard at the supply intake position.

All of the circuits are wired in flat twin and earth 70°C thermoplastic cable; all cables are concealed within the building structure and circuit protection is by BS EN 61009 RCBOs and BS EN 60898 circuit breakers.

The ring final circuits are wired in 2.5mm² twin and earth cable with a 1.5mm² CPC.

The copper water installation pipework and the incoming oil line are bonded to the main earthing terminal.

Using the information given, answer the following:

1 The installation needs to be isolated before any testing can be carried out. List the steps required to ensure safe isolation at the origin.

2 State the human senses which may be used to identify a loose connection.

3 List the three documents that should be completed on the completion of an initial verification for this installation.

4 List five items which must be included on the schedule to be fixed adjacent to the distribution board.

5 State five checks which would be carried out within a single phase distribution board and detail what is being checked.

6 One of the ring circuits is 63 metres long. Describe in detail how the circuit should be tested and show all measured resistance values expected for each stage of the test. Assume the circuit is isolated.

7 Calculate the expected Z_s value for the ring circuit.

8 Describe in detail the expected pattern of results when the earth fault loop impedance is checked at each socket outlet by direct measurement.

9 Describe in detail how an insulation resistance test should be carried out on a ring final circuit (assume that the circuit is isolated).

Questions

1 An insulation resistance test has been carried out on a 6 way consumer unit. The circuits recorded values of 5.6 MΩ, 8.7 MΩ, >200 MΩ, >200 MΩ, 12 MΩ and 7 MΩ.

Calculate the total resistance of the installation and state giving reasons whether or not the installation is acceptable.

2 A ring circuit is 54 metres long and is wired in 2.5mm^2/1.5mm^2 thermoplastic cable. The protective device is a 32A BS EN 60898 type C device and the Z_e for the installation is 0.24Ω. The resistance of 2.5mm^2 copper is 7.41mΩ per metre and 1.5mm^2 copper is 12.1mΩ per metre.

Calculate:

(i)   The Z_s for the circuit

(ii) Will the protective device be suitable?

3 An A2 radial circuit is wired in 4mm^2 thermoplastic twin and earth cable; it is 23 metres long, the circuit has on it 3 twin 13 amp socket outlets. Protection is by a 30 amp BS 3036 semi enclosed fuse. Z_e for the installation is 0.6Ω.

Socket 1 is 12.5 metres from the consumer unit, socket 2 is 6 metres from socket 1, and socket 3 is 4.5 metres from socket 2.

Calculate:

(i)   The $R_1 + R_2$ value at each socket outlet

(ii) If the circuit protective device will be suitable.

4 A 9.5 kW electric shower has been installed using $10mm^2/4mm^2$ thermoplastic twin and earth cable which is 14.75 metres long. The circuit is to be connected to a spare way in the existing consumer unit. Protection is by a BS 3036 semi enclosed 45A rewirable fuse.

Z_e for the system is 0.7Ω.

The temperature at the time of testing is 20°C.

(i) Calculate $R_1 + R_2$ for this circuit.

(ii) Will this circuit meet the required disconnection time?

5 A ring final circuit is wired using $4mm^2$ singles contained in conduit, the circuit is 87 metres in length and is protected by a 32A BS 3871 type 2 circuit breaker. The maximum permissible Z_s taken from the on-site guide is 0.79Ω and the actual Z_e is 0.52Ω.

Calculate:

(i) The expected Z_s.

(ii) The maximum permissible length that could be allowed for a spur in $4mm^2$ cable.

6 List the certification that would be required after the installation of a new lighting circuit.

7 List three non-statutory documents relating to electrical installation testing.

8 List four reasons why an electrical installation condition report would be required.

9 Apart from a new installation, under which circumstances would a periodic inspection *not* be required?

10 (i) A ring circuit is wired in $2.5mm^2/1.5mm^2$. The resistance of the phase and neutral loops were each measured at 0.3Ω. Calculate the resistance between L and N at each socket after all interconnections have been made.

(ii) Calculate the end to end resistance of the CPC.

(iii) Calculate the resistance between L and CPC at each socket after all interconnections have been made.

11 A spur has been added to the ring circuit in question 10. The additional length of cable used is 5.8 metres.

Calculate $R_1 + R_2$ for this circuit.

12 What is a 'statutory' document?

13 What is a 'non-statutory document'?

14 Why is it important to carry out testing on a new installation in the correct sequence?

15 How many special installation and locations are listed in the BS 7671 amended to 2015?

16 State the effect that increasing the length of a conductor could have on its insulation resistance.

17 An installation has seven circuits. Circuits 1, 4 and 6 have insulation resistances of greater than 200 MΩ. Circuits 2, 3, 5 and 7 have resistance values of 50, 80, 60 and 50 respectively. Calculate the total resistance of the circuit.

18 State the correct sequence of tests for a new domestic installation, connected to a TT supply.

19 List in the correct sequence the instruments required to carry out the tests in question 18.

20 State the values of the test currents required when testing a 30mA RCD used for additional protection.

21 How many times the rated operating current is required to operate a type B BS EN 60898 circuit breaker instantaneously?

22 What is the maximum resistance permitted for a conductor used for main protective bonding?

23 What would be the resistance of 22 metres of a single 10mm^2 copper conductor?

24 Which type of supply system uses an earth electrode and the mass of earth for its earth fault return path?

25 Table 12.3 shows the resistance values recorded at each socket on a ring circuit during a ring circuit test after the interconnections had been made. Are the values as expected? If not, what could the problem be? (Temperature is 20°C.)
The end to end resistances of the conductors are line 0.45Ω, neutral 0.46Ω and CPC is 0.75Ω.

26 A lighting circuit is to be wired in 1mm^2 twin and earth thermoplastic cable; the circuit is protected by a 5 amp BS 3036 fuse. What would be the maximum length of cable permissible to comply with the earth fault loop impedance requirements (Z_s)? Z_e is 0.45.

Table 12.3 Ring circuit details

	L to N	L to CPC
Socket 1	0.225	0.35
Socket 2	No reading	No reading
Socket 3	0.224	No reading
Socket 4	No reading	0.35
Socket 5	0.34	0.50
Socket 6	0.4	0.35
Socket 7	0.22	0.35

27 With regard to the *On-site Guide* what are the stated earth loop impedance values outside of a consumer's installation for a TT, TNS, and TNCS supply?

28 A ring final circuit has twelve twin 13 amp socket outlets on it. How many unfused spurs would it be permissible to add to this circuit?

29 How many fused spurs would it be permissible to connect to the ring circuit in question 28?

30 Name the document which details the requirements for electrical test equipment.

31 State three extraneous conductive parts that could be found within a domestic installation.

32 State four exposed conductive parts commonly found within an electrical installation.

33 State the minimum CSA for a non-mechanically protected, supplementary bonding conductor that could used in a bathroom.

34 What is the minimum acceptable insulation resistance value permissible for a complete 400 volt a.c. 50 Hz installation?

35 State the test voltage and current required for an insulation test carried out on a 230 volt a.c. 50 Hz installation.

36 The Electricity at Work Regulations state that for a person to be competent when carrying out inspecting and testing on an electrical installation, they must be ………. what?

37 A 400 volt a.c. installation must be tested with an insulation resistance tester set at …….. volts.

38 List three requirements of GS38 for test leads.

39 List three requirements of GS38 for test probes.

40 Name a suitable piece of equipment that could be used for testing for the presence of voltage while carrying out the isolation procedure.

41 To comply with BS 7671 the purpose of the initial verification is to verify that……………….

42 To comply with GN 3, what are the four responsibilities of the inspector?

43 When testing a new installation, a fault is detected on a circuit. State the procedure that should be carried out.

44 State three reasons for carrying out a polarity test on a single phase installation.

45 On which type of ES lampholders is it *not* necessary to carry out a polarity test?

46 What is the minimum requirement of BS 7671 for ingress protection of electrical enclosures?

12 Exercises and questions

47 A 6 amp Type B circuit breaker trips each time an earth loop impedance test is carried out on its circuit. How could the Z_s value for this circuit be obtained?

48 What is the maximum rating permissible before a motor would require overload protection?

49 Identify the type of circuit breaker that should be used for:
   (a) discharge lighting in a factory
   (b) a large transformer
   (c) a three phase motor.

50 Identify three warning labels and notices that could be found in an installation.

51 The circuits in Table 12.4 have been tested and the earth fault loop impedance values for each circuit are as shown. Using the rule of thumb method identify whether the circuits will comply with BS 7671.

52 State the minimum IP rating for fixed equipment in Zone 2 of a bathroom.

53 State the minimum size for a main protective bonding conductor installed in a TNS system with 25mm^2 meter tails.

54 Identify the documentation that should be completed after the installation of a cooker circuit.

55 State four non-statutory documents.

56 State four statutory documents.

57 State the sequence of colours for a new three phase and neutral system.

58 An initial verification of an electrical installation is to be undertaken. State three conditions which must be verified during the inspection.

59 State two statutory documents to which the inspector may refer to during an inspection.

60 State two non-statutory documents to which the inspector may refer to during an inspection.

Table 12.4 Z_s values

Measured Z_s	Maximum Z_s
0.86Ω	1.2Ω
0.68Ω	0.96Ω
1.18Ω	1.5Ω
2.8Ω	4Ω
1.75Ω	2.4Ω

61 State the purpose of an initial verification as described in GN3.

62 State two supply parameters which can only be supplied by the DNO.

63 An insulation resistance test is to be carried out on a circuit which contains surge protected socket outlets. State the conductors to which the test is to be applied to and the minimum test voltage to be applied.

64 State three special locations which are divided into zones.

65 Calculate the maximum earth fault loop impedance permissible for a 30mA RCD used to protect a circuit connected to a TN-S supply.

66 State the two tests which must be carried out to confirm I_{pf}.

67 State two methods which can be used to verify voltage drop in a circuit.

68 Which type of inspection requires voltage drop to be verified?

69 State the test instruments which can be used to verify phase rotation.

70 State two points on three phase installation where phase rotation should be verified.

71 State three documents which on which the value of I_{pf} should be recorded.

72 State one point connected to the main earthing terminal by each of the following conductors:

(a) circuit protective

(b) main protective bonding

(c) earthing conductor.

73 List three special locations which may be found in a private dwelling.

74 State three possible outcomes of a circuit breaker interrupting a fault which is two times larger than its rated I_{cn}.

75 The measured Z_s value for a ring final circuit is 1.16Ω; the maximum permitted value provided by BS 7671 is 1.37Ω. Calculate using the rule of thumb whether or not the measured value will be acceptable.

Now for the exam

There is no doubt that the City and Guilds 2391 inspecting and testing exams and assessments are very difficult and it is quite right that they are.

The 2391-50 is for the initial verification of electrical installations and the 2391-51 is the qualification for periodic inspections; these qualifications can be taken in any order.

It is possible to do both awards at the same time. This award is the 2391-52 level 3 award in initial and periodic inspection. It requires the candidate to take a 60-question online exam and carry out a visual photo inspection along with a written 4-question assessment and practical assessment on a test rig. The practical will include some fault finding. During the practical assessment both an electrical installation certificate and a condition report have to be completed.

The 2391-50 consists of an online 40-question open book exam, and two practical assignment tasks. These will include an inspection and test on a test rig and 4 written questions. During the practical assessment an electrical installation certificate must be completed.

2391-51 consists of an online 40-question multiple choice exam along with a 30-minute photo exercise, a practical assessment and a 4-question written assessment. During the practical exercise a condition report is to be completed.

The initial verification of an installation or circuit is very important and it must be carried out correctly. The documentation serves two purposes.

First, it is a certificate which should be provided to show that the installation has been installed correctly, and also that it has been verified as being safe to use. This certification should not be issued until the installation has been tested and the results have been verified as being satisfactory.

It would be a pointless exercise to just inspect and test an electrical installation when it is completed. It is vital that the installation is inspected regularly during the installation before any cables are covered up. For that reason the initial verification must begin at day one and continue through to completion.

Remember when you sign the electrical installation certificate you are taking responsibility for every part of the installation that you have inspected and tested. The initial verification should not be taken lightly because if an accident occurs due to an unsafe installation, the person who has carried out the inspection and test may well be held responsible.

In many ways initial verification should be easier than carrying out a periodic inspection on an existing installation – the person carrying out the test should know their way around it, as they would have seen it as it was being installed. In many cases the inspector will also be the installer.

A periodic inspection and test is of course a different proposition. It is virtually impossible to carry out this type of inspection without having a lot of experience in the type of installation being inspected, as in many cases judgements have to be made. It may be that the installation being inspected does not comply with the current edition of the wiring regulations, would that mean that it is unsafe? It is for you to decide, you are being paid as a professional person and you should be offering your clients the best advice that it is possible to give. Whilst it is nice to have a qualification for the inspection and testing of electrical installations, the qualification alone will not make up for a lack of experience.

Any training course should be seen as a pleasurable learning experience, and it will be if you select the correct training centre. Before signing up to any course, check out the percentage pass rate for the centre in which you are intending to study. If you choose the best centre possible you will be taught why we inspect and test and how to interpret any results obtained while testing. This is very important; the knowledge gained will be of far more value to you than the certificate. If you fail an exam it will not prevent you from carrying out inspecting and testing providing you have learnt how and why. If you pass your exam but do not understand what you are doing then the inspection and test becomes a pointless and dangerous exercise.

A good training centre will provide you with all of the required information which will enable you to carry out inspecting and testing safely and efficiently; it will also provide you with some information on exam techniques.

Because of the way this exam is set up knowledge alone will not guarantee an exam pass. Unfortunately it is clear that although many candidates are very good at the practical side, they fail the exam although they are more than competent when it comes to inspecting and testing. The reason that this happens is that although there is an understanding of the subject, the candidates use the wrong terminology when answering the questions. The following section may help.

The best advice of all is that you must put in the study time; it is nearly impossible for you to pass this exam if you only study during lessons.

Here are a few tips for the exam.

- Arrive in good time to allow yourself to settle.
- Make sure that you have some form of photo ID.
- Check that you have a pen and a calculator before you leave home; if you have coloured pencils and a ruler so much the better but they are not vital.
- Make sure that your phone is turned off!
- The written part of the 2391 assessments are different for the 2391-50, 2391-51 and 2391-52.
- Read each question carefully before attempting to answer it.
- Don't panic, just take your time and answer the questions as accurately as you can.
- Answer each question clearly, do not write a lot of waffle and expect the examiner to sort out the bits they need – they won't!
- Answer each question fully and write down the obvious; the examiner will not accept that you know something unless you show them.
- Use bullet points to answer questions such as 'how would you carry out a test on a ring circuit'. It is much easier to write down each step as you would carry out the test.
- Answer the questions clearly and do not overcomplicate the answer.
- Show all calculations as you will get some marks even if you get the answer to the calculation wrong.
- If you are asked to name three items, just name three as only the first three will be counted. If you write any more you are just wasting time.

Use the correct terminology, such as:

- Insulation resistance tester (not Megger).
- Low resistance ohm meter (not continuity tester or milli ohm meter).
- Electrical installation certificate or EIC (not electrical installation report).
- Schedule of test results (not schedule of tests or schedule of results).
- Electricity at Work *Regulations* 1989 or EAWR (not Electricity at Work Act).
- Health and Safety at Work *Act* 1974 or HASAWA (not Health and Safety at Work Regulations).
- Electrical safety quality and continuity regulation or ESQCR.

If describing RCD tests always provide the current value, do not just put times $\frac{1}{2}$ half, times 1 and times 5.

If answering questions about protective bonding always use the word protective; it could be main protective bonding or protective supplementary bonding.

Always use the correct values:

- M for megohm
- m for milliohm
- ms for milliseconds
- mA for milli amp
- I_{pf} not PFC
- The value of I_{pf} should be provided in kA.

The City and Guilds 2391/2400 inspection and test exams are based on information given in Guidance Note 3, BS 7671 wiring regulations, the *On-site Guide* and GS38. It is important that any person intending to take this exam has a sound knowledge of these documents.

The written paper will require in-depth answers; for example the question could ask how you would carry out an insulation test on a domestic lighting circuit. It is easier to answer the question using bullet points as shown than it is to write an essay.

- Ensure safety whilst test is being carried out.
- Gain permission to isolate the circuit
- Isolate circuit to be tested using the safe isolation procedure.
- Secure the test area by putting up barriers and signs.
- Remove all lamps.
- If the removal of lamps is difficult or fluorescent/transformers are on the circuit open the switches controlling them.
- Isolate or bypass all electronic equipment and equipment that may be damaged or give false readings.
- Link out any electronic switches.
- Use an insulation resistance tester set on 500 volts d.c.
- Ensure instrument is accurate and operate the instrument with the leads together and then apart to check correct operation.
- If it is necessary, disconnect the live conductors
- Do not disconnect the CPC from the earth terminal
- Test circuit between live conductors.
- Operate any two way switches controlling points where lamps have been removed.

- Join live conductors together and test between them and earth.
- Operate all two way switches.

The insulation resistance should be above 2MΩ. If it is less than this, further investigation must be carried out as a latent defect may exist.

- An insulation resistance value of 1MΩ is acceptable in some cases.
- Replace all lamps and remove any shorting links used.
- Replace all covers.
- Remove barriers and signs.
- Leave circuit safe and operational.

In any exam it is vital that the question is read carefully; often it is better to read a question several times to try to understand what is being asked.

Make sure that you show all calculations even if the question does not require you to. It is a good habit to get into and often you will get marks for showing the correct calculation even if the answer is wrong.

If the question asks for a fully labelled diagram, then marks are awarded for the diagram and the labelling. Where the question asks 'explain with the aid of a diagram' then a diagram and a written explanation are required.

When the question asks for a list then you will be expected to list a sequence of events in the correct order – just a list without any explanation. If you are required to state something then a statement is required in no particular order.

Example 1

List the sequence of dead tests.

(a) Continuity of protective bonding conductors and circuit protective conductors.

(b) Ring final circuit test.

(c) Insulation resistance test.

(d) Polarity.

State three statutory documents relating the inspecting and testing of electrical installations.

- The Electricity at Work Regulations 1989
- The Health and Safety at Work Act 1974
- The Electricity Supply Regulations.

Remember:

Always try to answer the questions in full using the correct terminology, for example: If asked 'which is the type of inspection to be carried out on a new installation', the answer must be: an initial verification. For the document required for moving a switch or adding a socket, the answer must be: An electrical installation minor works certificate, *not* just minor works.

Do not waste time copying out the question and write as clearly as you can.

CHAPTER 13

Answers

Exercise 1

1

Megger.

Certificate No: 2

ELECTRICAL INSTALLATION CERTIFICATE

(REQUIREMENTS FOR ELECTRICAL INSTALLATIONS BS7671 [IET WIRING REGULATIONS])

DETAILS OF THE CLIENT

Client: MRS VOLKS

Address: As BELOW.

INSTALLATION ADDRESS

Occupier:

Address: 3 Wagon Close, Somerset
SO3 6HT

DESCRIPTION AND EXTENT OF THE INSTALLATION

Description of Installation

SHOWER CIRCUIT IN DOMESTIC INSTALLATION

Extent of installation covered by this Certificate:

SHOWER CIRCUIT IN BATHROOM

(use continuation sheet if necessary)   see continuation sheet No:

(Tick boxes as appropriate)

New installation ☐

Addition to an existing installation ☑

Alteration to an existing installation ☐

FOR DESIGN, CONSTRUCTION, INSPECTION AND TESTING

I being the person responsible for the Design, Construction, Inspection & Testing of the electrical installation (as indicated by my signature below), particulars of which are described above, having exercised reasonable skill and care when carrying out that Design, Construction, Inspection & Testing, hereby CERTIFY that the design work for which I have been responsible is to the best of my knowledge and belief in accordance with BS7671: 2008 amended to 2015 except for any departures, if any, detailed as follows.

Details of departures from BS7671 as amended (Regulations 120.3 and 133.5):

NONE

Details of permitted exceptions (Regulation 411.3.3). Where applicable, a suitable risk assessment(s) must be attached to this certificate.

NONE

Risk assessment attached ☐

The extent of liability of the signatory is limited to the work described above as the subject of this Certificate.

Name (IN BLOCK LETTERS)   Date: Today

Company: Megger
Address: Archcliffe Road
Dover
Kent

CT17 9EN

Signature:

Tel No: 01223 767667

NEXT INSPECTION

I the designer, recommend that this installation is further inspected and tested after an interval of not more than 10 YEARS

Page 1 of 6

This form was developed by Megger Limited and is based on the model shown
in Appendix 6 of BS7671 : 2015   © Megger Limited 2015

2

SUPPLY CHARACTERISTICS AND EARTHING ARRANGEMENTS	(Tick boxes and enter details, as appropriate)

Earthing
TN-C
TN-S ✓
TN-C-S
TT
IT

Other source of supply
(to be detailed on
attached schedules)

Number and Type of Live Conductors

a.c.　　　　　d.c.

1-Phase,2-Wire ✓　2-wire

2-Phase,3-Wire　　3-wire

3-Phase,3-Wire　　Other

3-Phase,4-Wire

Confirmation of supply polarity ✓

Nature of Supply Parameters

Nominal voltage, U/U_0 [(1)]　　230 V

Nominal frequency, f [(1)]　　50 Hz

Prospective fault current, I_{pf} [(2)]　　0.3 kA

External loop impedance, Z_e [(2)]　　0.24 Ω

(Note: (1) by enquiry, (2) by enquiry or by measurement)

Supply Protective Device Characteristics

BS (EN) 1361

Type 11

Rated Current　100　A

PARTICULARS OF INSTALLATION REFERRED TO IN THE CERTIFICATE	(Tick boxes and enter details, as appropriate)

Means of Earthing
Distributor's Facility ✓

Installation Earth Electrode

Maximum Demand
Maximum demand (load)　85　kVA/Amps (Delete as appropriate)

Details of installation Earth Electrode: (where applicable)
Type: (e.g. rod(s), tape etc)　N/A　Location:　N/A　　Electrode resistance to earth:　N/A Ω

Main Protective Conductors

Earthing Conductor:　　material CU　csa 16 mm² Connection / Continuity verified ✓

Main protective bonding conductors:　material CU　csa 10 mm² Connection / Continuity verified ✓

To water installation pipes ✓	To gas installation pipes ✓	To oil installation pipes	To structural steel
To lightning protection	To other　Specify		

Main Switch / Switch-Fuse / Circuit-Breaker / RCD

Location: Garage
BS, Type: 61008-1
No of poles: 2

Current rating: 100　A

Fuse / device rating or setting:　—　A

Voltage rating: 230　V

If RCD main switch
Rated residual operating current $I_{\Delta n}$　30　mA

Rated time delay　N/A　ms

Measured operating time (at $I_{\Delta n}$)　60　ms

COMMENTS ON EXISTING INSTALLATION	(in the case of an alteration or additions see Section 633)

Appears to be in good condition.

SCHEDULES

The attached schedules are part of this document and this Certificate is valid only when they are attached to it.
　　1　Schedules of Inspections and　1　Schedules of Test Results are attached.
(Enter quantities of schedules attached)

This form was developed by Megger Limited and is based on the model shown
in Appendix 6 of BS7671 : 2015　© Megger Limited 2015

3

GENERIC SCHEDULE OF TEST RESULTS

Certificate No: 2

DB reference no I
Location Garage
Zs at DB Ω 0.24
I$_{pf}$ at DB (kA)
Correct supply polarity confirmed
Phase sequence confirmed (where appropriate)

Details of circuits and/or installed equipment vulnerable to damage when testing
RCD PROTECTION

Details of test instruments used (state serial and/or asset numbers)
Continuity — 081446
Insulation resistance — 081446
Earth fault loop impedance — 076H90
RCD
Earth electrode resistance

Tested by:
Name (Capitals)

Signature   Date

Circuit Number	Circuit Description	Circuit Details — Overcurrent device BS(EN)	type	rating (A)	breaking capacity (kA)	Conductor details Reference Method	Live (mm2)	cpc (mm2)	Ring final circuit continuity Ω r1 (line)	m (neutral)	r2 (cpc)	Continuity Ω (R1 + R2) or R2 R1 + R2 *	R2	Insulation Resistance Insulation (MΩ) Live - Live	Live - Earth	Polarity ✓ or	Insert ✓ or	Zs Ω	RCD @ I$_{\Delta n}$ (ms)	@ 5I$_{\Delta n}$ (ms)	Test button operation	Remarks (continue on a seperate sheet if necessary)
9	SHOWER	60898	B	50	6	100	10	4	1	1	1	0.24	1	>200	>200	✓	✓	0.39	60	17	✓	

* Where there are no spurs connected to a ring final circuit this value is also the (R1 + R2) of the circuit

4

The maximum value of Z_s for a 50A type B circuit breaker given in BS 7671 is 0.87Ω to correct for temperature using the rule of thumb. We must multiply the maximum Z_s by 0.8:

$$0.87 \times 0.8 = 0.69\Omega$$

As the measured value of Z_s is lower than the corrected value it will comply with the requirements of BS7671.

The measured value of Z_s is lower than the calculated $Z_e + R_1 + R_2$ value due to the measured value being taken when the system is live. The system will have bonding conductors connected and possibly other items providing parallel paths for the test currents and this will lower the measured Z_s value.

5

MINOR ELECTRICAL INSTALLATION WORKS CERTIFICATE

To be used only for minor electrical works which do not include the provision of a new circuit

Part 1 : Description of the minor work carried out

1. Details of the client...........Mrs Volks......................... Date work completed......Today........

2. Installation address or location...........3 Wagon Close, Bath, Somerset SO3 6HT...........................

3. Description of the minor work carried out....................Replace existing shower unit...

4. Details of any departures from BS 7671 : 2018 for the circuit worked on. See regulation 120.3. 133.1.3 and 133.5 where applicable a suitable risk assessment must be completed and attached to this document.

    Risk assessment attached ☐   No Departures

5. Comments on the existing installation (including any defects observed. See regulation 644.1.2)

    ....Good condition, visual inspection only...

Part 2: Presence and suitability of the installation earthing and bonding arrangements (see regulation 132.16)

1. System earthing arrangement : TT ☐   TN-S ☑   TN-C-S ☐

2. Earth fault loop impedance (Z_{db}) at the distribution board supplying the final circuit...0.39Ω

3. Presence of adequate main protective conductors:

Earthing conductor   ☑

Main protective bonding conductor(s) to: Water ☑   Gas ☑   Oil ☐ Structural steel☐ Other.........................☐

Part 3: Circuit details

DB reference No:...1.... DB location and type ......Garage..............

Circuit No:......9......... Circuit description............Shower circuit flat PVC twin and earth...

Conductor sizes: Live conductors...10....mm^2 CPC......4...mm^2

Circuit overcurrent protective device: BS(EN)...60898 Type...B..Rating.........50.....A

Part 4: Test results for the altered or extended circuit (where relevant and practicable)

Protective conductor continuity: R_2 .... Ω   or $R_1 + R_2$...0.24.........Ω

Continuity of ring final circuit conductors: L/L............Ω   N/N..............Ω   CPC/CPC..............Ω

Insulation resistance: Live - Live...>200...MΩ   Live - Earth...>200..MΩ

Polarity correct: ☑   Maximum measured earth fault loop impedance: Z_s...0.39...Ω

RCD operation: ☐   Rated residule operating current ($I_{\Delta n}$)......30...mA

Disconnection time ......60....ms   Test button operation is satisfactory   ☑

Part 5: Declaration

I certify that the work covered by this certificate does not impair the safety of the existing installation and that the work has been designed, constructed, inspected and tested to the best of my knowledge and belief, at the time of my inspection complied with BS 7671 except as detailed in Part 1 of this document.

Name...	Signature..
For and on behalf of..	Position..
Address..	Date..

6

(When answering these types of question, I use bullet points where possible as it is easier to add item in if you miss any.)

- Use an earth loop impedance tester with a no trip setting.
- Ensure that the tester is working correctly, is not damaged and is accurate.
- Check that test leads comply with GS38.
- Ensure that the shower unit is live.
- Place the earth test lead onto the earth point of the shower unit using a crocodile clip.
- Place the neutral test lead onto the N connection of the shower and the line test lead onto the line connection. (*If a 2 lead tester is used omit the neutral connection.*)
- Press test button and wait for measured value to show.
- Remove leads, line first.
- Isolate shower and fit cover.
- Record measured value onto test result schedule or minor works certificate.
- Leave circuit safe.

Exercise 2

1

Megger.

ELECTRICAL INSTALLATION CERTIFICATE

(REQUIREMENTS FOR ELECTRICAL INSTALLATIONS BS7671 [IET WIRING REGULATIONS])

DETAILS OF THE CLIENT
Client: MR CHEWTER
Address: 2 STONEPOUND ROAD, HURSTPIERPOINT, W. SX BN6Y33

INSTALLATION ADDRESS
Occupier:
Address: AS ABOVE

DESCRIPTION AND EXTENT OF THE INSTALLATION
Description of Installation
DWELLING WIRED IN TWIN AND EARTH PVC CABLE

(Tick boxes as appropriate)

New installation ☐

Addition to an existing installation ☑

Alteration to an existing installation ☐

Extent of installation covered by this Certificate:
NEW RING FINAL CIRCUIT CONSISTING of
7 SOCKET OUTLETS. WITH RCD MAIN SWITCH

(use continuation sheet if necessary)   see continuation sheet No:

FOR DESIGN, CONSTRUCTION, INSPECTION AND TESTING
I being the person responsible for the Design, Construction, Inspection & Testing of the electrical installation (as indicated by my signature below), particulars of which are described above, having exercised reasonable skill and care when carrying out that Design, Construction, Inspection & Testing, hereby CERTIFY that the design work for which I have been responsible is to the best of my knowledge and belief in accordance with BS7671: 2008 amended to ToDAY   except for any departures, if any, detailed as follows.
Details of departures from BS7671 as amended (Regulations 120.3 and 133.5):

NONE

Details of permitted exceptions (Regulation 411.3.3). Where applicable, a suitable risk assessment(s) must be attached to this certificate.

NONE

Risk assessment attached ☐

The extent of liability of the signatory is limited to the work described above as the subject of this Certificate.
Name (IN BLOCK LETTERS)   Date:
Company: Megger
Address: Archcliffe Road
            Dover   Signature:
            Kent

            CT17 9EN
                                       Tel No:

NEXT INSPECTION
I the designer, recommend that this installation is further inspected and tested after an interval of not more than   10 YRS

Page 1 of 6

SUPPLY CHARACTERISTICS AND EARTHING ARRANGEMENTS *(Tick boxes and enter details, as appropriate)*

Earthing
TN-C
TN-S
(TN-C-S)
TT
IT

Other source of supply
(to be detailed on
attached schedules)

Number and Type of Live Conductors
(a.c.)   d.c.
1-Phase,2-Wire   (2-wire)
2-Phase,3-Wire   3-wire
3-Phase,3-Wire   Other
3-Phase,4-Wire

Confirmation of supply polarity

Nature of Supply Parameters
Nominal voltage, U/U$_0$ (1)   230   V
Nominal frequency, f (1)   50   Hz
Prospective fault current, I$_{pf}$ (2)   1·45   kA
External loop impedance, Z$_e$ (2)   0·18   Ω
(Note: (1) by enquiry, (2) by enquiry or by measurement)

Supply Protective Device Characteristics
BS (EN) 1361
Type 2
Rated Current 100 A

PARTICULARS OF INSTALLATION REFERRED TO IN THE CERTIFICATE   *(Tick boxes and enter details, as appropriate)*

Means of Earthing
Distributor's Facility ✓
Installation Earth Electrode

Maximum Demand
Maximum demand (load)   88   kVA/Amps   *(Delete as appropriate)*

Details of installation Earth Electrode: *(where applicable)*
Type: *(e.g. rod(s), tape etc)*   Location: N/A   Electrode resistance to earth: Ω

Main Protective Conductors

Earthing Conductor: material CU   csa 16   mm² Connection / Continuity verified ✓

Main protective bonding conductors: material Cu   csa 10   mm² Connection / Continuity verified ✓

| To water installation pipes ✓ | To gas installation pipes N/A | To oil installation pipes ✓ | To structural steel N/A |
| To lightning protection N/A | To other N/A   Specify | | |

Main Switch / Switch-Fuse / Circuit-Breaker / RCD

Location: HALL CUPBOARD
BS, Type: 610081
No of poles: 2

Current rating: 100 A
Fuse / device rating or setting: — A
Voltage rating: 230 V

If RCD main switch
Rated residual operating current I$_{\Delta n}$ 30 mA
Rated time delay NONE ms
Measured operating time (at I$_{\Delta n}$) 34 ms

COMMENTS ON EXISTING INSTALLATION   *(in the case of an alteration or additions see Section 633)*

THERE IS NO DOCUMENTATION FOR THE EXISTING INSTALLATION
PROTECTION FOR ALL CIRCUITS IS BY BS 3036 REWIRABLE
FUSES.
THE INSTALLATION APPEARS TO BE IN GOOD CONDITION
ALTHOUGH I WOULD ADVISE THAT A PERIODIC INSPECTION IS
CARRIED OUT.

SCHEDULES

The attached schedules are part of this document and this Certificate is valid only when they are attached to it.
Schedules of Inspections and | Schedules of Test Results are attached. |
(Enter quantities of schedules attached)

GENERIC SCHEDULE OF TEST RESULTS

Certificate No: 2

DB reference no	1
Location	WALL CUPBOARD
Zs at DB Ω	0.18
Ipf at DB (kA)	
Correct supply polarity confirmed	✓
Phase sequence confirmed (where appropriate)	

Details of circuits and/or installed equipment vulnerable to damage when testing

RCD MAIN SWITCH

Tested by:

Name (Capitals)

Signature

Date

Details of test instruments used (state serial and/or asset numbers)

Continuity	CJK 1047
Insulation resistance	'' ''
Earth fault loop impedance	'' ''
RCD	
Earth electrode resistance	N/A

Test results

Circuit Number	Circuit Description	Overcurrent device BS(EN)	type	rating (A)	breaking capacity (kA)	Reference Method	Live (mm2)	cpc (mm2)	Ring final circuit continuity Ω r1 (line)	rn (neutral)	r2 (cpc)	Continuity Ω (R1 + R2) or R2 R1 + R2 *	R2	Insulation Resistance Insulation (MΩ) Live - Live	Live - Earth	Polarity Insert ✓ or	Zs Ω	RCD @ I∆n (ms)	@ 5I∆n (ms)	Test button operation	Remarks (continue on a seperate sheet if necessary)
7	Ring final circuit	3036	2	30	2	100	4	1.5	0.24	0.24	0.4	0.4	1	>200	200	✓	0.58	46	34	✓	

* Where there are no spurs connected to a ring final circuit this value is also the (R1 + R2) of the circuit

2

MINOR ELECTRICAL INSTALLATION WORKS CERTIFICATE

To be used only for minor electrical works which do not include the provision of a new circuit

---

Part 1 : Description of the minor work carried out

1. Details of the client...............Mr Chewter................................ Date work completed.........Today.....

2. Installation address or location.................2 Stonepound Road...Hurstpierpoint...

3. Description of the minor work carried out.........Twin 13a socket outlet added to existing ring final circuit

4. Details of any departures from BS 7671 : 2018 for the circuit worked on. See regulation 120.3. 133.1.3 and 133.5 where applicable a suitable risk assessment must be completed and attached to this document.

    Risk assessment attached | N/A |

5. Comments on the existing installation (including any defects observed. See regulation 644.1.2)

    Earthing and bonding up to date but would advise a periodic inspection is carried out.....................................

Part 2: Presence and suitability of the installation earthing and bonding arrangements (see regulation 132.16)

1. System earthing arrangement : TT ☐   TN-S ☐   TN-C-S ☑

2. Earth fault loop impedance (Z_{db}) at the distribution board supplying the final circuit...0.71...Ω

3. Presence of adequate main protective conductors:

Earthing conductor   ☑

Main protective bonding conductor(s) to: Water ☑   Gas n/a   Oil ☑   Structural steel n/a   Other.........N/A..... ☐

Part 3: Circuit details

DB reference No:...1.   **DB location and type** ....Hall cupboard single phase consumer unit.................

Circuit No:......7......... **Circuit description**..............Ring final circuit...

Conductor sizes: Live conductors......4....mm^2 CPC......1.5.........mm^2

Circuit overcurrent protective device: BS(EN)...3036 Type......2.....**Rating**......30.....A

Part 4: Test results for the altered or extended circuit (where relevant and practicable)

Protective conductor continuity: R_2 .... Ω   or $R_1 + R_2$...0.53......Ω

Continuity of ring final circuit conductors: L/L...0.24.Ω   N/N...0.24.Ω   CPC/CPC 0.64...Ω

Insulation resistance: Live - Live...>200..MΩ   Live - Earth...>200 MΩ

Polarity correct: ☑   Maximum measured earth fault loop impedance: Z_s ...0.71...Ω

RCD operation: ☑   Rated residule operating current ($I_{\Delta n}$)...30...mA

Disconnection time ......46.......ms   Test button operation is satisfactory   ☑

Part 5: Declaration

I certify that the work covered by this certificate does not impair the safety of the existing installation and that the work has been designed, constructed, inspected and tested to the best of my knowledge and belief, at the time of my inspection complied with BS 7671 except as detailed in Part 1 of this document.

Name...	Signature..
For and on behalf of...	Position..
Address..	Date..

Exercise 3

1

Circuit	$R_1 + R_2$ W	Actual Z_s W	Max permitted	Pass/Fail
1	$\dfrac{10.40 \times 23}{1000} = 0.24\Omega$	$0.63 + 0.24 = 0.87\ \Omega$	$1.09 \times 0.8 = 0.872\ \Omega$	Pass
2	$\dfrac{20.51 \times 48}{4000} = 0.234\Omega$	$0.63 + 0.234 = 0.86\ \Omega$	$1.37 \times 0.8 = 1.09\ \Omega$	Pass
3	$\dfrac{16.71 \times 53}{4000} = 0.22\Omega$	$0.63 + 0.22 = 0.85\ \Omega$	$1.37 \times 0.8 = 1.09\ \Omega$	Pass
4	$\dfrac{30.20 \times 12}{1000} = 0.36\Omega$	$0.63 + 0.36 = 0.99\ \Omega$	$2.73 \times 0.8 = 2.18\ \Omega$	Pass
5	$\dfrac{30.20 \times 38}{1000} = 1.14\Omega$	$0.63 + 1.14 = 1.77\ \Omega$	$3.64 \times 0.8 = 2.9\ \Omega$	Pass
6	$\dfrac{36.20 \times 57}{1000} = 2.06\Omega$	$0.63 + 2.06 = 2.69\ \Omega$	$7.28 \times 0.8 = 5.82\ \Omega$	Pass

2

SECTION A. DETAILS OF THE CLIENT / PERSON ORDERING THE REPORT

Name *MARY MINT (MRS)*

Address *POLO HOUSE   MINTO CLOSE   NUTLEY   KENT KT3 8JL*

SECTION B. REASON FOR PRODUCING THIS REPORT

CLIENT REQUEST

Date(s) on which inspection and testing was carried out *TODAY*

SECTION C. DETAILS OF THE INSTALLATION WHICH IS THE SUBJECT OF THIS REPORT

Occupier *MARY MINT*

Address *24 TREBOR ROAD   DOVER   KENT   KH10 1DR*

Description of premises (tick as appropriate)

Domestic [✓]   Commercial [ ]   Industrial [ ]   Other (include brief description) [ ]

Estimated age of wiring system *18* years

Evidence of additions / alterations *YES*   If yes, estimate age *5* years

Installation records available? (Regulation 621.1) *YES*   Date of last inspection *2010*   (date)

SECTION D. EXTENT AND LIMITATIONS OF INSPECTION AND TESTING

Extent of the electrical installation covered by this report

COMPLETE FIXED INSTALLATION

Agreed limitations including the reasons (see Regulation 634.2) *NONE OTHER THAN THOSE PRINTED BELOW*

Agreed with: *N/A*

Operational limitations including the reasons (see page no   ) *NONE*

The inspection and testing detailed in this report and accompanying schedules have been carried out in accordance with BS 7671:2008 (IET Wiring Regulations) as amended to *3 : 2015*

It should be noted that cables concealed within trunking and conduits, under floors, in roof spaces, and generally within the fabric of the building or underground, have **not** been inspected unless specifically agreed between the client and inspector prior to the inspection. An inspection should be made within an accessible roof space housing other electrical equipment.

SECTION E. SUMMARY OF THE CONDITION OF THE INSTALLATION

General condition of the installation (in terms of electrical safety) *INSTALLATION IN GOOD CONDITION WITH THE EXCEPTION OF THE EARTHING CONDUCTOR WHICH SHOULD BE INCREASED TO 16mm² WITH NO EVIDENCE OF SUPPLIMENTARY BONDING*

Overall assessment of the installation in terms of its suitability for continued use

*An unsatisfactory assessment indicates that dangerous (code C1) and/or potentially dangerous (code C2) conditions have been identified

SECTION F. RECOMMENDATIONS

Where the overall assessment of the suitability of the installation for continued use above is stated as UNSATISFACTORY, I/We recommend that any observations classified as 'Danger present' (code C1) or 'Potentially dangerous' (code C2) are acted upon as a matter of urgency. Investigation without delay is recommended for observations identified as 'further investigation required' (code FI). Observations classified as 'Improvement recommended' (code C3) should be given due consideration.

Subject to the necessary remedial action being taken, I/We recommend that the installation is further inspected and tested by *2026*

SECTION G. DECLARATION

I/We, being the person(s) responsible for the inspection and testing of the electrical installation (as indicated by my/our signatures below), particulars of which are described above, having exercised reasonable skill and care when carrying out the inspection and testing, hereby declare that the information in this report, including the observations and the attached schedules, provides an accurate assessment of the condition of the electrical installation taking into account the stated extent and limitations in section D of this report.

Inspected and tested by:	Report authorised for issue by:
Name (Capitals) *AN OTHER*	Name (Capitals) *BN OTHER*
Signature	Signature
For/on behalf of *AN ELECTRICAL*	For/on behalf of *AN ELECTIRCAL*
Position *SITTING DOWN*	Position
Address *2 MINT LANE NUTLEY KENT*	Address
Date *TODAY*	Date

SECTION H. SCHEDULE(S)

schedule(s) of inspection and *1* schedule(s) of test results are attached. *1*

The attached schedule(s) are part of this document and this report is valid only when they are attached to it.

Certificate No: 12

SECTION I. SUPPLY CHARACTERISTICS AND EARTHING ARRANGEMENTS | Tick boxes and enter details, as appropriate.

Earthing arrangements	Number and Type of Live Conductors		Nature of supply Parameters	Supply Protective Device
TN-C ☐	- a.c. ☑	d.c. ☐	Nominal voltage, U/Uo (1) 230 V	BS (EN) 1361
TN-S ☑	1-phase, 2-wire ☑	2-wire ☐	Nominal frequency, f (1) 50 Hz	
TN-C-S ☐	2-phase, 3-wire ☐	3-wire ☐	Prospective fault current Ipf (2) 1-2 kA	Type 2
TT ☐	3-phase, 3-wire ☐	Other ☐	External loop impedance, Ze (2) 0·63 Ω	
IT ☐	3-phase, 4-wire ☐		Note: (1) by enquiry	Rated current 80 A
	Confirmation of supply polarity ☐		(2) by enquiry or by measurement	

Other sources of supply (as detailed on attached schedule) ☐

SECTION J. PARTICULARS OF INSTALLATION REFERRED TO IN THE REPORT | Tick boxes and enter details as appropriate

Means of Earthing | **Details of Installation Earth Electrode** (where applicable)

Distributor's facility ☑ | Type
Installation earth electrode ☐ | Location N/A
| Resistance to Earth Ω

Main Protective Conductors

Earthing conductor	Material Cu	csa 6 mm²	Connection / continuity verified ☑
Main protective bonding conductors	Material Cu	csa 10 mm²	Connection / continuity verified ☑

To water installation pipes ☐	To gas installation pipes ☐	To oil installation pipes ☐	To structural steel ☐
To lightning protection ☐	To other ☐	Specify NOT SEEN	

Main Switch / Switch-Fuse / Circuit-Breaker / RCD

Location UNDERSTAIRS	Current rating 100 A	If RCD main switch	
BS(EN) 5713	Fuse / device rating or setting N/A A	Rated residual operating current (I∆n) N/A m/
No of poles 2	Voltage rating 240 V	Rated time delay ms
		Measured operating time (at I∆n) ms

SECTION K. OBSERVATIONS

Referring to the attached schedules of inspection and test results, and subject to the limitations specified at the *Extent and limitations of inspection* and testing section.

No remedial action is required ☐   The following observations are made ☑ (see below)

OBSERVATION(S) Include schedule reference, as appropriate	CLASSIFICATION CODE
EARTHING CONDUCTOR NOT LARGE ENOUGH	C2
NO EVIDENCE OF PROTECTIVE SUPPLIMENTARY BONDING	C2
WOULD ADVISE RCD PROTECTION INSTALLED FOR ALL CIRCUITS TO BRING THE INSTALLATION UP TO THE REQUIREMENTS OF BS7671	C3

One of the following codes, as appropriate, has been allocated to each of the observations made above to indicate to the person(s) responsible for the installation the degree of urgency for remedial action.
C1 - Danger present. Risk of injury. Immediate remedial action required
C2 - Potentially dangerous - urgent remedial action required
C3 - Improvement recommended
FI - Further investigation required without delay

DB reference no 1
Location Understairs
Zs at DB Ω 0.63
Ipf at DB (kA)
Correct supply polarity confirmed ✓
Phase sequence confirmed (where appropriate)

Details of circuits and/or installed equipment vulnerable to damage when testing
PIR To OUTSIDE LTs
SHAVER SOCKET MAY CAUSE Low
IR VALUES

Details of test instruments used (state serial and/or asset numbers)
Continuity: CJK 1047
Insulation resistance:
Earth fault loop impedance: Multifunction
RCD: N/A
Earth electrode resistance: N/A

Tested by:
Name (Capitals):
Signature:
Date: Today

Test results

Circuit Number	Circuit Description	BS(EN)	type	rating (A)	breaking capacity (kA)	Reference Method	Live (mm2)	cpc (mm2)	r1 (line)	rn (neutral)	r2 (cpc)	R1 + R2 *	R2	Live–Live	Live–Earth	Polarity Insert ✓ or	Zs Ω	@ IΔn	@ 5IΔn	Test button operation	Remarks
1	COOKER	60898	B	40	6	100	6	2.5	/	/	/	0.24	/	>200	>200	✓	0.87	N/A	N/A	N/A	
2	RING 1	"	B	32	6	100	2.5	1.5	0.35	0.35	0.23	0.13	/	>200	>200	✓	0.86	"	"	"	
3	RING 2	"	B	32	6	100	4.0	1.5	0.24	0.24	0.58	0.12	/	>200	>200	✓	0.85	"	"	"	
4	IMMERSION	"	B	16	6	100	1.5	1.0	/	/	/	0.36	/	>200	>200	✓	0.99	"	"	"	
5	LIGHTING	"	C	6	6	100	1.5	1.0	/	/	/	0.12	/	>200	>200	✓	0.75	"	"	"	
6	LIGHTING	"	B	6	6	100	1.0	1.0	/	/	/	2.06	/	>200	>200	✓	2.69	"	"	"	

Exercise 4

GENERIC SCHEDULE OF TEST RESULTS

Certificate No: 6

DB reference no	
Location	
Zs at DB Ω 0·4	
Ipf at DB (kA)	
Correct supply polarity confirmed	
Phase sequence confirmed (where appropriate)	

Details of circuits and/or installed equipment vulnerable to damage when testing

Details of test instruments used (state serial and/or asset numbers)

Continuity	
Insulation resistance	
Earth fault loop impedance	
RCD	
Earth electrode resistance	

Tested by:
Name (Capitals)
Signature _____ Date _____

Test results

Circuit Number	Circuit Description	BS(EN)	type	rating (A)	breaking capacity (kA)	Reference Method	Live (mm2)	cpc (mm2)	r1 (line)	rn (neutral)	r2 (cpc)	R1 + R2 *	R2	Live - Live	Live - Earth	Polarity Insert ✓ or	Zs Ω	@IΔn	@5IΔn	Test button operation	Remarks
1		B	40	10	100	6·0	2·5	/	/	/	0·15	/				0·55				
2		B	32	10	100	4·0	1·5	6·33	0·33	0·37	6·3	/				0·7				
3		B	32	10	100	2·5	1·5	0·50	0·50	0·32	0·7	/				1·1				
4		B	16	10	100	2·5	1·5	/	/	/	0·53	/				0·73				
5		B	6	10	100	1·5	1·0	/	/	/	0·96	/				1·36				
6		B	6	10	100	1·0	1·0	/	/	/	1·55	/				1·95				

Column groups: **Circuit Details** — Overcurrent device; **Conductor details**; **Ring final circuit continuity Ω**; **Continuity Ω (R1 + R2) or R2**; **Insulation Resistance Insulation (MΩ)**; **Polarity**; **Zs Ω**; **RCD (ms)**.

* Where there are no spurs connected to a ring final circuit this value is also the (R1 + R2) of the circuit

Exercise 5

1 An earth fault loop impedance tester.
2 GS38.
3 Ensure the complete system is isolated.

   Disconnect the earthing conductor from the main earthing terminal.

   Check that the test equipment is not damaged, is accurate and working correctly.

   Connect the green test lead to the disconnected earthing conductor and the brown and blue leads to the line and neutral terminals at the live side of the main switch.

   Push the test button and record the measured value as Z_e.

   Disconnect test leads, reconnect earthing conductor, replace any covers and leave safe.

4 $\dfrac{50}{0.03} = 1667\Omega$

5 Although the maximum value for a TT installation it should be noted that this value would be considered as unstable; action should be taken to reduce any measured value to 200Ω or less.
6 An earth electrode resistance test could be carried out.
7 An earth electrode resistance tester.
8 Two additional earth electrodes should be used.

Exercise 6

1 Gain permission to isolate the installation.

   Using an approved voltage indicator to GS38, ensure voltage indicator is working.

   Switch off the main switch and lock off using the correct lock off device and put up isolation notice.

   Use the voltage indicator to test between line conductors, then all line conductors to neutral, then all line conductors and neutral to earth.

   Re-check that the voltage indicator on a known supply or a proving unit to ensure that it is still working correctly.

2 Touch by pulling on cable. Sight to see conductor not fully in terminal. Smell where arcing has taken place.
3 Electrical installation certificate, schedule of inspections and a schedule of test results.
4 The type and composition of each circuit.

   The method used for fault protection.

   Type of protective device.

Rating of protective device for each circuit.

Any circuit or equipment which would be vulnerable to electrical tests.

5　Correct colour ID of conductors: brown for line, blue for neutral, green and yellow for CPC.

Manual operation of circuit breakers and RCDs.

Correct rating of protective devices.

Conductors not damaged.

Correct barriers in place.

Single pole protective devices in line conductors only.

Protection against electromagnetic effects where cables enter ferromagnetic enclosures.

Conductors in correct sequence.

Circuits correctly identified.

Protection against mechanical damage where conductors enter the enclosure.

6　Use a low resistance ohm meter with leads nulled.

Check that the instrument is not damaged, is accurate and working correctly.

Disconnect all ring final circuit conductors at the consumer unit.

Test line conductor end to end. $\dfrac{7.41 \times 63}{1000} = 0.466\Omega$.

Neutral conductor will be the same.

Test CPC end to end $\dfrac{12.1 \times 63}{1000} = 0.762\Omega$.

Record all readings on to the schedule of test results.

Cross connect the line and N conductors and test between L and N at each socket outlet.

$\dfrac{0.466 + 0.466}{4} = 0.23\Omega$ would be the expected measured value

although a small difference of, say, 0.05Ω would be acceptable.

Cross connect the CPC and line conductors and measure between line and earth at each socket outlets $\dfrac{0.466 + 0.762}{4} = 0.65\Omega$ should

be the measured value at each socket, although due to the difference in conductor size the measured value will be slightly less for sockets nearer the consumer unit and increase towards the centre of the ring. The highest value should be entered as $R_1 + R_2$.

Reconnect all conductors and replace any covers as required.

7　$Z_s = Z_c + (R_1 + R_2)$ 0.12 + 0.65 = 0.77Ω

8　The earth loop impedance resistance will be less on socket outlets near the consumer unit and will increase towards the centre of the ring.

9 Use an insulation resistance tester set to 500 volts d.c. with leads and probes to GS38.

Ensure correct operation of the tester by testing with leads together and then apart.

Check tester and leads for damage and accuracy.

Unplug any equipment which is on ring and switch off any fused connection units and any fixed equipment which may cause false readings.

Test between all live conductors.

Test between all live conductors and earth, ensuring that the CPC is connected to the main earthing terminal while test is carried out.

Reconnect any conductors which have been disconnected and replace any barriers.

Record values onto the schedule of test results.

Questions

Q1 Any values which are indicated as ≥ (greater than) can be disregarded as the true value is unknown.

$$\frac{1}{5.6} + \frac{1}{8.7} + \frac{1}{12} + \frac{1}{7} = 0.51 \qquad \frac{1}{0.51} = 1.92M\Omega$$

Enter into calculator as:

$$5.6 \ x^{-1} + 8.7 \ x^{-1} + 12 \ x^{-1} + 7 \ x^{-1} = x^{-1} = 1.92$$

This is acceptable as the total insulation resistance is greater than $0.5M\Omega$ and each circuit is greater than $2M\Omega$

Q2 $2.5mm^2/1.5mm^2$ cable has a resistance of $19.51m\Omega$ per metre. The resistance of 54 metres is:

$$\frac{54 \times 19.51}{1000} = 1.05\Omega \qquad \frac{1.05}{4} = 0.26\Omega$$

$R_1 + R_2$ for the circuit is 0.26Ω

$Z_s = Z_e + R_1 + R_2 \qquad 0.24 + 0.26 = 0.5\Omega$

From Table 41.3 in BS 7671 the maximum value for a type C 32A device is 0.68 use the rule of thumb to compensate for conductor operating temperature and ambient temperature. $0.68 \times 0.8 = 0.54\Omega$

Therefore this circuit will comply.

Q3 From Table I1 in the *On-site Guide* the $r_1 + r_2$ value for the $4mm^2/1.5mm^2$ conductors is $16.71m\Omega$ /M. The $R_1 + R_2$ value at the sockets is

Socket 1. $\qquad \frac{12 \times 16.71}{1000} = 0.2\Omega$

Socket 2. $\dfrac{18 \times 16.71}{1000} = 0.3\Omega$

Socket 3. $\dfrac{20.5 \times 16.71}{1000} = 0.34\Omega$

Actual Z_s will be $0.34 + 0.6 = 0.94\Omega$. The maximum Z_s for a 30A BS 3036 fuse from table 41.2 in BS7671 (0.4s) is given as 1.04Ω.

Using the rule of thumb for temperature correction the maximum permissible value is: $1.04 \times 0.8 = 0.83\Omega$

This value is NOT acceptable as the actual Z_s is higher than the maximum

Q4 From Table I1 in the *On-site Guide* the $r_1 + r_2$ value for 10mm²/4mm² copper is 6.44mΩ/M.

$$R_1 + R_2 = \dfrac{6.44 \times 14.75}{1000} = 0.095\Omega$$

Z_s for the circuit is $0.095 + 0.7 = 0.795\ (0.8)\Omega$

The maximum Z_s for a 45A BS 3036 fuse from table 41.4 in BS 7671 (5s) is 1.51Ω.

Corrected for temperature. $1.51 \times 0.8 = 1.20\Omega$. As the actual Z_s is lower the circuit will comply.

Q5 (i) From Table I1 in the *On-site Guide* 4mm² copper conductors have a resistance of 4.61mΩ/M.

$r_1 + r_2$ for the phase and CPC both in 4mm² is 9.22mΩ/M

$$R_1 + R_2 = \dfrac{9.22 \times 67}{1000} = 0.61\Omega \quad \text{As it is a ring} \quad \dfrac{0.61}{4} = 0.51\Omega$$

The $R_1 + R_2$ value is 0.15Ω

$Z_s = Z_e + R_1 + R_2.\ 0.63 + 0.15 = 0.78\Omega$ as the maximum permissible is $1\Omega \times 0.8 = 0.8\Omega$ this value is acceptable.

(ii) As the maximum permissible Z_s is given as 0.9 and is taken from the *On-site Guide* no correction for temperature is required. We must subtract the Z_e from the actual Z_s to find the maximum permissible $R_1 + R_2$ value.

$$R_1 + R_2 = 0.9 - 0.63 = 0.27\Omega$$

We must now subtract the actual $R_1 + R_2$ value from the maximum permissible value.

$$0.27 - 0.15 = 0.12\ \Omega$$

The maximum resistance that our spur could have is 0.12Ω. To calculate the length we must transpose the calculation :

$$\dfrac{mv \times \text{length}}{1000} = R$$

to find the total length transpose to $\dfrac{R \times 1000}{mV}$

Therefore the length $= \dfrac{0.12 \times 1000}{9.22} = 13m$

Total length of cable for the spur will be 13 metres.

Q6 (a) Electrical installation certificate

(b) Schedule of test results

(c) Schedule of inspections

Q7 BS 7671 Wiring regulations

On-site Guide

Guidance Note 3

GS38

Q8 Due date

Client request

Change of use

Change of ownership

Before alterations are carried out

After damage such as fire or overloading

Insurance purposes

Q9 Where there is recorded regular maintenance

Q10 (i) L-N are 0.3Ω each. The total loop will be 0.6 and the L-N at each socket after interconnection will be $\dfrac{0.6}{4} \times 0.15$ ohms

(ii) Cpc must be $0.3 \times 1.67 = 0.5\Omega$

(iii) $R_1 + R_2$ loop will be $0.5 + 0.3 = 0.8\Omega$. After interconnection the $R_1 + R_2$ value at each socket on the ring will be $\dfrac{0.8}{4} = 0.2\Omega$

Q11 $2.5\text{mm}^2/1.5\text{mm}^2$ has a resistance of $19.51\text{m}\Omega$ per metre.

5.8 metres of the cable will have a resistance of:

$$\frac{5.8 \times 19.51}{1000} = 0.113\Omega$$

0.113 is the resistance of the additional cable. $R_1 + R_2$ for this circuit will now be $0.2 + 0.113 = 0.313\Omega$

Q12 A statutory document is a legal requirement.

Q13 A non-statutory document is a recommendation/guidance.

Q14 Safety, as the satisfactory dead tests ensure that the installation is safe to energise. It will also avoid having to repeat tests if one of the tests is unsatisfactory.

Q15 There are 19 special installations and installations

Q16 The insulation resistance would decrease

Q17 $\dfrac{1}{50} + \dfrac{1}{80} + \dfrac{1}{60} + \dfrac{1}{50} = 0.069 \quad R = \dfrac{1}{0.069} = 14.45\text{M}\Omega$

Q18 Continuity of protective bonding and CPCs

Ring final circuit test

Insulation resistance test

Polarity

Live polarity at supply
Earth electrode (Z_e)
Prospective fault current
Residual current device
Functional tests

Q19 Low resistance ohm meter
Low resistance ohm meter
Insulation resistance tester
Low resistance ohm meter
Approved voltage indicator
Earth loop impedance tester
Prospective short circuit current tester
RCD tester

Q20 15mA ($\times \frac{1}{2}$). 30mA ($\times$1). 150mA ($\times$5) (*only 150mA if used for*

supplementary protection)

Q21 5 times

Q22 0.05Ω

Q23 From Table 11 in the *On-site Guide* 10mm^2 copper has a

resistance of 1.83mΩ per metre. $\dfrac{1.83 \times 22}{1000} = 0.4\Omega$

Q24 TT system

Q25 Socket 1. Good circuit
Socket 2. CPC and N reversed polarity
Socket 3. P and N reversed polarity
Socket 4. P and CPC reversed polarity
Socket 5. Spur
Socket 6. Loose connection of N
Socket 7. Good circuit

Q26 From Table 41.2 Z_s for a 5A BS 3036 fuse is 9.10Ω.
The $r_1 + r_2$ value for 1mm^2 copper from Table I1 is 36.2mΩ.
Maximum resistance permissible for the cable:
9.10 – 0.45 = 8.65Ω

Maximum length of circuit is $\dfrac{8.65 \times 1000}{36.2} = 238.95$m

(*Problem with volt drop if the circuit was this long*)

Q27 TT 21Ω
TNS 0.8Ω
TNCS 0.35Ω

Q28 12 socket outlets, one for each socket on the ring.

Q29 Unlimited number

Q30 GS38

Q31 Taps. Radiators. Steel bath. Water and gas pipes, etc.

Q32 Steel conduit and trunking. Metal switch plates and sockets. Motor case, etc.

Q33 $4mm^2$

Q34 $1M\Omega$

Q35 500V 1mA

Q36 In possession of technical knowledge or experience or suitably supervised.

Q37 500 Volts d.c.

Q38 Flexible. Long enough but not so long that they would be clumsy. Insulated. Identified. Suitable for the current.

Q39 Finger guards. Fused. Maximum 4mm exposed tips. Identified.

Q40 Approved voltage indicator or test lamp.

Q41 Fixed equipment complies with British Standards, all parts correctly selected and erected, not damaged.

Q42 To ensure no danger to persons and livestock and that no damage occurs to property.

To compare the results with the design criteria.

Take a view on the condition of the installation and advise on any remedial works required.

In the event of a dangerous situation, to make an immediate recommendation to the client to isolate the defective part.

Q43 Ensure the fault is repaired and retest any parts of the Installation which test results may have been affected by the fault.

Q44 To ensure that all single pole switches are in the line conductor.

Protective devices are in the phase conductor.

ES lampholders are correctly connected.

The correct connection of equipment

Q45 E14 and E27 as they are all insulated.

Q46 The top surface must comply with IP4X. The sides and front IP2X or IPXXB.

Q47 By calculation $Z_s = Z_e + R_1 + R_2$. Or use a low current earth fault loop test instrument.

Q48 0.37kW regulation 552.1.2

Q49 (a) Type C

(b) Type D

(c) Type C

Q50 Safety electrical connection do not remove.

Voltage in excess of 230 volts where not expected.

Notice for RCD testing.

Where isolation is not possible by the use of a single device.

Where different nominal voltages exist.

Periodic test date.

Warning non-standard colours

Dual isolation notice

Q51 $1.2 \times 0.8 = 0.96\Omega$

$0.96 \times 0.8 = 0.76\Omega$

$1.5 \times 0.8 = 1.2\Omega$

$4 \times 0.8 = 3.2\Omega$

$2.4 \times 0.8 = 1.92\Omega$

Q52 IPX4

Q53 $10mm^2$

Q54 Electrical installation certificate

Schedule of test results

Schedule of inspection

Q55 BS 7671. Electrical wiring regulations.

On-site Guide

GS38

Guidance Note 3

Q56 Health and Safety at Work Act 1974

Electricity Supply Regulations

Electricity at Work Regulations 1989

Construction Design and Management Regulations

Building Regulation Part P.

(Appendix 2 of BS 7671 covers statutory regulations)

Q57 Brown (L1), Black (L2), Grey (L3) and Blue (N).

Q58 (a) The installation is compliant with the requirements of Section 511 (all to a BS).

(b) Correctly selected and erected to BS 7671 and to manufacturers instructions.

(c) Not visibly damaged or defective so as to impair safety.

Q59 (a) EAWR 1989

(b) HASAWA 1974

(c) ESQCR 2002

Q60 (a) BS 7671

(b) *On-site Guide*

(c) GS38

Q61 That the installation complies with the design and construction aspects of BS 7671 as far as reasonably practicable.

Q62 Nominal voltage and nominal frequency.

Q63 Live conductors and live conductors to earth 250 volts d.c.

Q64 Rooms containing a bath or shower. (bathroom or shower will be ok but will only count as one mark)

Q65 $\dfrac{230}{0.03} = 7666\Omega$ (if a TT system use 50 volts but must use 230 for TNS and TNCS)

Q66 Prospective short circuit current and prospective earth fault current.

Q67 (a) Calculation using the resistance of the cable

(b) using voltage drop tables.

Q68 A Periodic inspection.

Q69 (a) Rotating disc type.

(b) indicator lamps.

Q70 (a) At the origin

(b) Each distribution board

(c) Isolators

(d) Motor starters, (not motors)

Q71 (a) Electrical installation certificate

(b) Electrical installation condition report

(c) Schedule of test results

Q72 (a) Earth terminal of an electrical accessory and the MET

(b) Extraneous conductive parts and MET (Water, Gas, Oil installation)

Q73 (a) Room containing a bath or shower

(b) Sauna

(c) Swimming pool

(d) Underfloor heating

(e) Photovoltaic system

Q74 (a) The device will interrupt the supply but will become unserviceable and need to be replaced.

(b) The contacts will weld together and the device will not interrupt the supply.

(c) The device will disintegrate causing damage to equipment around it.

(d) The device may cause a fire.

Q75 $1.37 \times 0.8 = 1.096\Omega$ as the measured value of 1.6 is higher than the calculated maximum this circuit will be unacceptable.

Glossary

Ambient temperature	The temperature of the air or environment where equipment is to be used
Appliance	Any item of current using equipment other than an electric motor or luminaire
ADS	Automatic disconnection of supply, disconnection of supply under fault conditions
Barrier	Protection against unintentional contact with live parts from any usual direction
Basic protection	Protection from electric shock under no fault conditions
Bonding conductor	A protective conductor used to provide equipotential bonding
BS 7671	British Standard for the requirements for electrical installations
Cartridge fuse	A fuse element enclosed in a cartridge
Circuit breaker	A device which is manufactured to interrupt overload and fault currents automatically
Circuit protective conductor	A conductor used to connect exposed conductive equipment to the main earthing terminal
Class I equipment	Equipment which has a means of connection to the earthing system and does not only rely on basic insulation
Class II equipment	Equipment which has basic and supplementary insulation but does not have a provision for connection to earth
Class III equipment	Equipment in which protection against electric shock is reliant on extra low voltage (SELV)
Competent person	A person who has sufficient skills, technical knowledge and experience to undertake a job safely
Conduit	Part of an enclosed wiring system which allows cables to be drawn in and replaced
Connector	A coupler with female contacts which is intended to be attached to a flexible cable remote from the supply

Consumer's unit	A type of distribution board containing a main switch and protective devices (fuses or circuit breakers) normally used in domestic installations
CPC	Circuit protective conductor
CSA	Cross sectional area
Current using equipment	Equipment which converts electrical energy into another form of energy such as heat, light and electric motors
Danger	Risk of injury
Design current	The current which a circuit is designed to carry under the conditions in which it is installed
Disconnector	A device for breaking circuits not under load also known as an **Isolator**
Distribution board	An enclosure containing protective devices, sometimes called a consumer's unit or fuse board
Distribution circuit	A circuit supplying a distribution board/consumer's unit
Double insulated	Equipment which is protected by both basic and supplementary insulation in the event of a fault between live conductors and earth
Double insulation	Insulation which has both basic and supplementary insulation
Double pole switch	A mechanical device which is capable of making and breaking under normal load conditions the line and neutral conductors of a circuit
Earth electrode	Metal conductive part which is buried in soil or other conductive material which is in contact with earth
Earth fault loop impedance	The resistance of the earth fault loop, this is known as impedance and is shown as Z_s
Earth fault loop path	The path which the fault current should flow in the event of a circuit fault to earth
Earthing	Connection of the exposed conductive parts of an installation to the main earthing terminal of the installation
Earthing conductor	A conductor connecting the main earthing terminal to the means of the system earthing
Electric shock	A dangerous physiological effect caused by a current passing through a human body or livestock
Electrical equipment	Any item which is used as part of the electrical installation, including test instruments
Electrical installation	Electrical equipment installed to provide energy for a specific purpose

Electrical installation certificate	Certificate issued on completion of a new installation, a new part of an existing installation, or an alteration to an existing installation
Emergency switching	A system comprising of one operation to cut off the electrical supply in the event of an emergency
Enclosure	An enclosure providing protection of electrical equipment against the required external influences
Equipotential bonding	An electrical connection which is installed to maintain extraneous and exposed conductive parts at the same potential
ESC	Electrical Safety Council
Exposed conductive part	A conductive part of equipment which can be touched, is not normally live but could become live in the event of a fault
External influence	Anything which is not part of the electrical system but may have an effect on it
Extraneous conductive part	A part of an electrical installation which is not normally live but may become live in the event of a fault. The metal cases of a washing machine or central heating pump for example
FELV	Functional extra low voltage. An extra low voltage system which does not meet the requirements of SELV or PELV and must meet all of the test requirements for low voltage circuits
Fixed equipment	Electrical equipment which is intended to be fixed / secured
Fixed wiring	The wiring forming the electrical installation up to and including the electrical outlets
Flexible cable	A cable which is in designed to be flexed while in service
Flexible wiring system	A wiring system which is designed to flex while in use without degradation
Functional switching	Switching on or off electrical energy to an installation or a circuit
Fuse	A device in which a wire or metal strip melts when a certain value of current passes through it
Fused connection unit	An accessory which contains a cartridge fuse and is used for the connection, and fusing down of equipment. It is often referred to as a fused spur
Impedance	Resistance in an AC circuit
Inspection	An electrical system being checked without testing

Insulation	Non conductive material surrounding or supporting a conductor
Isolation	Switching off an electrical circuit, part of a circuit or a complete installation for reasons of safety
Isolator	A mechanical switching device used to isolate an electrical system or part of an electrical system
Line conductor	A conductor used to carry electrical energy, this conductor is often incorrectly referred to as the live conductor. Both the line and neutral are live conductors. In a standard circuit the cables will consist of a line conductor, a neutral conductor and a protective conductor
Luminaire	Light fitting
Main earthing terminal (MET)	A bar used for the connection of earthing and bonding conductors. In a domestic installation this is usually found inside the consumers unit
MCS	Microgeneration Certification Scheme
Origin	The point at which the supply is connected to the installation
Overcurrent	A current which is greater than the current carrying capacity of the circuit conductors
Overload current	A larger current flowing in a circuit greater than the design current
Part P	The part of the building regulations which cover the requirements of electrical work carried out in domestic installations
PELV	Protective extra low voltage. An extra low voltage system which is not electrically separated from Earth, but satisfies all of the other requirements of SELV
PIR	Passive infra red detector
Point	The part of the fixed wiring which is intended for the connection of current using equipment
Prospective fault current (IPf)	The highest current which could flow in a system under fault conditions
Prospective short circuit current (PSCC)	The maximum current that could flow between phase and neutral on a single phase supply or between phase conductors on a three phase supply
RCBO	A device which is an RCD and also provides overcurrent protection
RCCB	Residual current circuit breaker

RCCBO	Residual current circuit breaker with overload protection
Residual current device (RCD)	A device which switches off when there is an imbalance of current between live conductors of the same circuit. This device does not provide overcurrent protection
Resistance	Opposition to the flow of current
Ring final circuit	A circuit in which the conductors form a loop starting at the supply point and returning to the same point
SELV	Separated extra low voltage. An extra low voltage system which is electrically separated from other systems
Short circuit current	The current which flows when two live conductors are touched together
Socket outlet	A device which is installed into the fixed wiring system to receive a plug
Spur	A cable connected to a ring or radial circuit to supply an additional point
Stationary equipment	A piece of electrical equipment which is fixed or has a mass of more than 18kg and is not provided with a carrying handle
STC	Standard test conditions
Switch	A mechanical device which is capable of making and breaking current under normal load conditions
Switch disconnector	A device which is used for switching/isolating a circuit under load
Switched disconnector	A switch device which when in the open position satisfies the requirements of an isolator
Testing	A process applied to an electrical installation or circuit requiring the use of specific measuring equipment to verify its condition
Voltage bands:	
Band I	Installations where shock protection is provided by the value of voltage, these circuits would include telecommunication, alarm and bell systems. Extra low voltage will be considered to be band I
Band II	Installations using low voltage are classed as band II circuits, low voltage is between 50v a.c. and 1000v a.c.0

Index

Note: 'F' after a page number indicates a figure; 't' indicates a table; 'g' indicates a glossary entry.

Electrical Inspection, Testing and Certification:

A Guide to Passing the City & Guilds 2391 Exams

Second Edition

Michael Drury

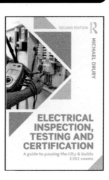

- Assists students in decoding questions and terminology in the UK's City & Guilds new 2391 exam structure
- Addresses the areas which had previously been highlighted as poor in the Chief Examiner's feedback of candidate performance
- Uses realistic exam questions as examples, explaining *why* and *how* a correct answer has been selected, rather than simply providing an answer

An essential guide to the City & Guilds 2391-50 and 51: Initial Verification and Certification of Electrical Installation and Periodic Inspection and Testing, also C&G 2391-52: an amalgamation of Initial Verification and Periodic Inspection and Testing of electrical installations.

There is a full coverage of technical and legal terminology used in the theory exams; including the structure of exam questions and their interpretation. By running through examples of realistic exam questions in a step-by-step fashion, this book explains how to decode the questions to achieve the most suitable response from the multiple-choice answers given.

This book is ideal for all electricians, regardless of their experience, who need a testing qualification in order to take the next step in their career.

24 April 2018 | 172 pages
Paperback: 978-0-8153-7799-3